AF522417

MODERN TEACHING OF CHEMISTRY

MODERN TEACHING OF CHEMISTRY

[Strictly According to the UGC Syllabus for B.Ed. Course]

By
MAN PAL SINGH

ANMOL PUBLICATIONS PVT. LTD.
NEW DELHI - 110 002 (INDIA)

ANMOL PUBLICATIONS PVT. LTD.
4374/4B, Ansari Road, Daryaganj
New Delhi - 110 002
Ph.: 23261597, 23278000
Visit us at: www.anmolpublications.com

Modern Teaching of Chemistry

First Published, 2004

ISBN 81-261-2074-6

PRINTED IN INDIA

Published by J.L. Kumar for Anmol Publications Pvt. Ltd., New Delhi - 110 002 and Printed at Mehra Offset Press, Delhi.

Contents

Preface

Education is a vast discipline and Teachers' Training is a vital part of it. The responsibilities of the educationists and educators are focused on the task of providing better training to the future teachers for their better learning and proper development. Needless to say that this responsibility can only be exercised, if the trainers are equipped with the required knowledge of the subject concerned. That's why it becomes essential for making adequate provisions for each course to the student-teachers or teacher trainees. The present series is designed for providing a solid workable base for all course-papers. It has been prepared strictly according to the syllabus of the B.Ed class, prescribed by the UGC for different universities.

No doubt, there are so many other books on the subject, available in the market, written by worthy authors. However, every writer has his or her own style and way of presentation. The present work also has its own features and characteristics.

In preparation of this series of texts, the editor had to refer to the works of other authors and information sources. The editor feels a deep sense of gratitude for incorporating their ideas in the text. Hopefully, this series would serve as a 'ready to refer' tool for all teachers, teacher-students and others.

—Editor

1

Introduction

A reasonable short definition of the scope of chemistry has been given as, "chemistry is the integrated study of the preparation, properties, structure and reactions of the chemical elements and their compounds and of the systems which they form."

Interpreted in the broader sense, this definition would include most of natural science, a consequence of the considerable areas of overlap which chemistry has with physical, biological, earth and material sciences.

Chemists tend to work on molecular rather than atomic systems, and on molecular structures and transformations rather than on phenomenon associated with simple substances only.

Chemical science is dynamic in scope and many chemists find themselves working in areas currently described as interfaces (*e.g.* molecular biology, solid state physics).

The Organisation

The traditional substructure of physical, organic, inorganic, and some times analytical chemistry has origin in the past activities of chemists and still continues to determine how chemistry is taught. Another possible sub-division would be the following: The structure and physical properties of pure substances, chemical transformations and applications of chemistry to complex systems. Application of chemistry would include such fields of current activity as molecular biology, material science and geochemical phenomenon.

Evolution and Development

The scientific method emerged in the 16th century with the realisation that investigation must supplement logic and intution in probing nature. This view point which owes its initiation to scientists like Bacon, Boyle, Galileo, Hooke, Newton and others, recorded its early successes in astronomy and mechanics. The method came to be gradually applied to chemical behaviour of matter, thanks to the pioneering efforts of Antoine Lavoisier in the late 18th century and this marked the birth of chemical science as we know it today.

Importance of Chemistry

Chemistry as considered an important subject in school curriculum as many professional and applied courses, directly or indirectly use the knowledge of chemistry. Moreover, the present age is the era of science and more number of people are being employed in scientific pursuits which require knowledge of chemistry.

Chemistry education is also necessary because of its immense value in the students' individual life as well as in society.

Chemistry is essentially a secondary school subject. At this level, it may be taught as a subject in its own right or as part of a broader science course identified by a variety of titles, *e.g.* integrated science, general science and modular science. The discipline may also feature as a component of courses in physical or biological sciences.

The most significant aspect of modern science is the impact it has had in solving a variety of problems of practical and technological importance as well as those related to the pressing problems of mankind. A large number of these problems require a proper understanding and application of chemical principles and processes.

The major threats to the present day civilisation are population explosion, hunger and disease, environmental pollution, depletion of sources of energy as well as natural resources. The growth of population is probably the greatest problem facing us.

In solving most of these pressing problems, chemists have a lot to do. Paracelus (1493-1541) said, 'the true use of chemistry is not to make gold, but to prepare medicines'. The problem of atmospheric pollution, if and when it is solved will only be done through an understanding of chemical dynamics.

Chemistry has made a significant contribution in the fields of drugs, fuels, agriculture, animal farming, fibres etc. In addition to these there are many other inter-disciplinary areas where the contribution of chemists is significant. In the area of *environmental pollution,* chemists arc finding better methods of analysis and solutions to get rid of pollution. There is the entire area of *marine chemistry* to investigate new sources for food and chemicals.

The involvement of chemists in real life problems has been pointed out in the previous section. In this section we take up a few specific areas in some details.

Drugs : Many a substances from natural sources have been used since times immemorial for treatment of diseases. For example, an extract from the bark of poplar, olive or willow trees was recommended more than two thousand years ago by Hippocrates—the father of medicine—for treating fever. At present we can isolate and purify the drug from natural sources and establish their chemical structure Sparsely occurring substances can be synthesized in the laboratory and in this way made available in abundance. More over even such drugs which do not occur in nature have also been obtained by synthesising them in the laboratory. *Aspirin* (Acetyl salicylic acid) is one of the earliest synthetic drug. *Salvarsan* was synthesized for treatment of syphilis. Some other prominent synthetic drugs are *Sulpha drugs, antibiotics, anaesthetics, anti-malarials* and wonder drug *cortisone*.

Food : Chemist have done a lot to increase food grain production and helped to bring about green revolution. *Synthetic fertilizers* were developed which provided the-essential elements for growth of plants. The use of these fertilizer led to higher yield of plants. *Insecticides, weed killers, fungicides* developed by chemists have contributed a lot to increase the availability of food grains for the mankind. In many countries farmers use laboratory made chemicals as *defoliants*. For example, magnesium chlorate when applied to ripened cotton crop causes the leaves to fall of thus making harvesting much cheaper and faster. Chemicals are also used in animal farming. For example, 'marlate'-a new insecticide-used as a dip or as a spray-kills' blood-sucking hornflies which attack cows. This step alone leads to 10% increase in milk supply.

Fuels : Till the middle of this century only naturally occurring substances such as wood, coal, coke etc. were used as fuels but now the situation has been completely changed by *processsing of petroleum Petrochemical industry* also provides many a useful chemicals like benzene, toluene, xylene, naphthalene etc. *Petrochemicals* provide the base of synthetic fibres, rubber, resins, detergents, refrigerants and explosives.

Fibres : Now a days we are producing *synthetic fibres* like *nylon, rayon* and *terylene*. These fibres in some respects excel the natural fibres. They are longer lasting, crease resistant and quick drying.

In addition to these there are many inter-disciplinary areas where contribution of chemicals is significant. For example production of glasses and ceramics, electronic, magnetic and optical materials; fibre-based composites; etc. In the area of *environmental pollution* chemists are finding better methods of analysis and solution to get rid of pollution. *Marine chemistry* is concerned with investigating new sources of food and chemicals.

Significance of Humanbeings

A nations most valuable resource is its people. The intelligence, creativity and talent that resides in the human mind awaits only its release and full development through education.

Chemistry in particular, is close to a nations' health and strength and to the well being of its people. Since chemistry touches the lives of every individual (through agriculture, industry, nutrition, industry, medicine, home environment etc.). We can easily say that an individuals' *every moment* is directly influenced by the understanding and therefore the utilization he or she can make of chemistry. Scientific discoveries, technological advances, the efficiency of work force, the exercising of citizens rights and quality of life are directly tied to the teaching of chemistry.

As a teacher of chemistry our goals should he

(i) to provide appropriate education in chemistry to everyone.

(ii) to fully develop human resources.

Having set our goal we should establish strategies to achieve the goals and should periodically evaluate our progress. We should always have an open mind to set even higher goals for some thing better,

For chemistry teaching in future the goals have to be set in a different light than in the past. This is due to the fact that in the past we aimed to educate a few scientists and engineers for our country. Now we aim at educating every one in basis of chemistry. With this in view students have to be classified on the basis of diversity in student interest and goals. The following five grasp can be easily identified.

(i) Those interested to become future chemists.

(ii) Those interested in other science-based professions (*e.g.* the biological and earth scientists, engineers, physicians, nutritionist etc.).

(iii) Those who are interested to become technical personnel. They will comprise the support system for science and technology.

(iv) Those who are likely to join industries, in health sciences and in agriculture.

(v) Ordinary citizens.

Every body needs sufficient knowledge of chemistry to function effectively in present day society. At present our society is being influenced by new drugs, synthetic materials, green revolu-tions in agriculture, micro-computers, micro-electronics etc.

In developing human resources we should aim at providing high quality education in chemistry. Such an education should provide opportunity for working in laboratory and for solving mathematical and intellectual problems. Students be encouraged to investigate, to explore, to use the library, to use the natural environment and to discuss chemical concepts and issues in order to provide them sufficient opportunity and experience to cope with benefits from products and processes of chemistry through out their lives.

Thus in future we shall have to give a much different chemistry curriculum which may either be in the form of a course suitable for every one at secondary level or of several streams for providing chemical education for every one. The syllabus for science students and non-science students has been discussed in chapter on curriculum.

2

History of the Subject

In our country Science is witnessed to be taught from very early days. If we look at the development of science in India, we find that very early in her civilisation India developed a great interest in mathematics and Ayurveda. During British Empire introduction of modern science in India was extremely slow, and the development of science in India was greatly accelerated after independence

Initial Growth

Before the rise of modern science in Europe, around the 17th century, the level of advancement in sciences (astronomy, mathematics, medicine, biology, metaphysics, etc) achieved by ancient and medieval societies in the old world did not differ appreciably from one cultural area to another. In India the development of sciences is as old as her civilisation itself. Her peculiar geographical position enabled her to become the natural meeting ground of many nations and cultures and, in consequence, played an important role in the transmission and diffusion of ideas very early in her civilisation, India developed medical and alchemical practices of the Ayurveda. In mathematics, India developed a great interest and aptitude and made notable contributions to the number theory, the decimal place value, algorism, trigonometry and algebra.

Modern Times

In the beginning of the nineteenth century people with varying degrees of scientific background (medical men, naturalists and engineers) started coming to India from Europe, at first in search of services under local princes and chieftains and later on to man the scientific surveys and establishments set up by the government. These men framed in European institutions and laboratories, spent the best part of their lives in India and left an excellent record of their work in various branches of science.

Despite such long contacts, introduction of modern science in India was extremely slow. There were many causes of this delayed reaction. The European scientific world in India was limited to field sciences and not to basic sciences that depended on mathematics and laboratory work. Indian fauna and flora attracted the attention of European naturalists from the seventeenth century. Modern zoological researchers in India had their beginnings in random and scattered observations by the naturalists on elephants, fishes, serpents, molluscs, birds and mammals.

Past Experiences

The Colonial Government adopted from the beginning a policy of secrecy and exclusion of Indians from Government scientific work. For technological jobs not considered of military importance, restrictions against the employment of Indians were relaxed as they could be employed very cheaply. As an example of the low salaries to trained and qualified Indians/ the order of the Governor-General dated 28th January 1835, establishing the

Calcutta Medical College, regulated that "as inducement for pupils of a respectable class to enter the Institution, the pay of the Native Doctors, who shall have been educated at the College, and have received the certificates of qualifications, shall be Rs. 30.00 p.m. rising to Rs. 50.00 p.m. after 14 years of service, whereas an European Assistant to the Superintendent of the college shall draw a staff salary of Rs. 600.00 p.m. in addition to his registered pay and allowances." Calcutta Medical College established in 1835 under a number of capable teachers, became an important institution, for the study of anatomy, physiology and medicine (along with surgery), as well as of chemistry, botany and natural philosophy.

First Research Institutions : Indians asked for Western Education in Science from the Government. The Colonial power agreed to create facilities but with miserly financial provision just sufficient to train a few clerks able to operate in English in Government offices and European Commercial houses, and not to enable the native people to pick up treasures from European science. So the hard way of working for science began. Mahendra Lal Sircar *(born* in 1833), an M.D. of the Calcutta Medical College, clearly saw that science would never develop and strike deep roots in this country under foreign tutelage and that Indians themselves must come forward to raise funds and found institutions for the training of scientists and organisation of research. In 1876 Dr. Sircar himself founded India's first research institution, "The Indian Association for the Cultivation of Science," completely under Indian Management and control with finances derived from private subscriptions. In the first half of the twentieth century the association developed into an important centre for research in the physical sciences - in optics, acoustics, scattering of light, X-rays and magnetism; and C.V. Raman performed his Nobel Prize Winning Experiments in its unassuming laboratories, *i.e.* western India. Jamshedji Tata a businessman, prepared plans for a similar institution for technical and scientific education and research which finally took the shape of "The Institute of Science" at Bangalore in the beginning of the twentieth century.

First College of Science : The Educational Despatch of 1854 paved the way for University education. In 1857 the three Universities of Calcutta, Bombay and Madras came into existence, but they came out just as examining bodies with powers to grant degrees. The actual teaching and academic work were left to the colleges. Fortified by the University Act of 1904, which empowered Universities to appoint professors and lecturers, to hold and manage educational endowments, and to erect, equip and maintain libraries, laboratories and museums, Asutosh Mookerji, Vice Chancellor of Calcutta University took the initiative in establishing the, first University College of Science. Undaunted by the Government's refusal to provide funds for the creation of professorships and other facilities, Asutosh obtained princely endowments from Sir Tarak Nath Palit and Sir Rash Behary Ghosh, who had amassed enormous amounts of money and property in the legal profession, and established a number of chairs in Chemistry, Physics, Applied Mathematics and Botany. According to the terms of the endowment, professorships could be filled only by Indians, a clause which further irritated the Government firmly entrenched in the view that only Europeans were suited for such high positions. This University College of Science, although starved financially all through, produced a group of physicists and chemists who received international recognition for their scientific developments and institutions staffed by high salaried Europeans and maintained and patronised by Government funds cut a sorry figure.

National Policy on Science

The development of science in India was greatly accelerated after independence (August 1947). In 1950 the Government of India appointed a Planning Commission for preparing a blueprint of all-round economic development. In 1954 the Indian Parliament accepted socialism as a political goal. Declaring these objectives, fullest emphasis was laid on the development of science and technology on all fronts. In 1957, the Government took one step further in adopting a National Science Policy Resolution that envisaged the cultivation of science and scientific research in all its aspects, assured an adequate supply, within the country, of research scientists of highest quality through an intensive programme of training, promised the availability of conditions and an atmosphere of academic freedom in which the creative talent of men and women would find full scope in scientific activity. The resolution thus reaffirmed the Government decision to encourage science and develop a healthy scientific community as a sound basis after a balanced economic development.

School Education

- In the beginning of twentieth century science was not a school subject in our country. The Report of the Secondary School Commission 1953, recommended the teaching of General Science as a compulsory subject in the high and higher secondary schools.

- The All India Seminar on the Teaching of Science in Secondary Schools (1956) dealt with almost all the problems facing the inclusion of General Science as a Core Subject for the Higher Secondary Classes - syllabus, apparatus, teaching aids, textbooks, science clubs, science museums, examination techniques etc. It suggested a uniform system of science teaching for the entire country, suited to its needs and resources.

- Indian Parliamentary and Scientific Committee (1962) studied the allied problems of science education in schools like

  — growth of school population,

  — shortage of qualified science teachers,

  — accelerated achievement in science,

  — demand for increase in technically trained manpower,

— growing importance of science in the affairs of mankind, and

— changes in the processes and goals of science.

- UNESCO Planning Commission (1963-64) worked on the problems of science education in India and suggested ways to improve it. As a follow up programme, Department of Science and Mathematics Education of the NCERT took up the pilot project of preparing new disciplinary science curricula at middle (VI-VIII) level text books (Physics, Chemistry, Biology), teachers' guides, science kits, kit guides, teacher training films and evaluation material. This Disciplined Science Programme was changed to Integrated Science Programme in mid-eighties.

- In 1970 under UNICEF Assisted Science Education Programme (SEP) Primary "Science is Doing" Programme was developed by NCERT, which was used in primary schools throughout the country. This Programme was a package of classes I and II Science Syllabus, Class III-V Science Texts, Teachers Guides, Primary Science Kit, Kit Guide, and Teacher Training Films. This Programme was changed to EVS Programme in mid-eighties.

- Kothari Commission (1964-66) recommendations were implemented in 1975 when Science for All (SFA) was introduced as a part of general education during the first ten years of schooling. With this 10+2+3 education scheme started with an additional year of schooling, in the country. First Disciplined Science Course (Physics, Chemistry, Biology) was introduced at secondary level (IX-X). This was changed to Disciplined Science A-Course (Physics, Chemistry, Biology) and Integrated Science B-Course. Students had an option either to take Science A-Course or to take Science B-Course. This again changed to just one Integrated Science Course for all. At +2, Senior Secondary level (XI-XII) Disciplined Science Course (3 different science subjects - Physics, Chemistry, Biology) started from the very beginning of 10+2+3 education scheme.

- Then in 1986 the National Education Policy Document (NPE-1986) came out (Chapter 23). If some one asks, "What is new in this new education policy (NPE-1986), "Implementation" perhaps will be the right answer. Much emphasis has been given to quality pre-service and in-service teacher education in the policy document. For this District Institutes of Education (DIETs) and Institutes of Advance Studies in Education (LASEs) were established throughout the country for Elementary and Secondary Education respectively.

New Trends

India is now engaged in a broad spectrum of scientific research, both fundamental and applied, in Government, Univer-sities and Private Research Establishments. In recent years there has been extraordinary success in developing new polymers, cera-mics, composites, superconductors, nanomaterials, smart materials and biomaterials. Biotechnology, Genetic Engineering and Biomedical Research are some other fields in which India has started entering.

AIDS (Acquired Immuno Deficiency Syndrome) poses a threat to India as a large number of people are infected with HIV (Human Immunodeficiency Virus). There are no drugs today for AIDS. The problem which has dogged anti HIV drugs is that resistant mutant forms of virus are formed within mere weeks. Lot of research work is to be done in search of a vaccine against HIV.

Per capita consumption ɔf energy in India is very low compared to developed countries, and even that we are unable to afford. In the years to come when we have already entered twenty-first century, we will need much more energy. The major effort in next decade would have to be through an increase in the production of coal and a search of new reserves of oil. We should also give due emphasis to new technologies for solar energy and hydrogen energy. India is also to expand its nuclear power programmes. Nuclear energy could play an important role in meeting our bulk energy requirement. Our expertise in the field of nuclear power technology as well as related research areas, is an asset for us.

For all this we need huge amounts of funds, as scientific research has become costly. If we compare the funds available for scientific research in India compared to some other developing countries, India has a very gloomy picture. South Korea had planned to increase its R&D spending to over 5 per cent of its GDP by year 2000. China had planned to increase its R&D investment from 0.5 per cent to 1.5 per cent of GDP by the year 2000. Unfortunately India's R&D expenditure has come down to 0.89 per cent of GDP from 1.1 per cent earlier. We need to realise that in order to be able to do competitive scientific research and development we have to bring our R&D expenditure to at least about 1.5 per cent of GDP and focus our efforts on a few selected programmes and projects.

Our scientists want people to support them and understand their needs, and they are to inform people about science, why science is needed, what they are doing and why. They (our scientists) are to build up partnership with public, industries, politicians and bureaucrats. They need informed friends of science at all places.

When changes which affect our future, are happening and will happen at such a rapid rate, and are based on science and technology, it is necessary that our scientists be more close to decision making.

In the existing state of scientific advancement and development of resources for research, the rate of scientific growth of developed countries is likely to continue to be faster than that of the developing countries like India. Science has become deeply involved with defence and big industry, with the result that big sciences like atomic energy, space, etc., has been and probably will remain concentrated in super powers, In nuclear, space, computer and few other sciences, developing nations like India are already at the mercy of super powers. To be self-sufficient we are to change our science curriculum right from the school stage. We are not just to teach science but also scientific method. We are not just to teach science content but also science processes, the ways in which scientists advance their knowledge and solve problems. Science should be presented to students as a way in which they can conduct an inquiry into the nature of things as well as a body of information built up by other people. The science processes are being neglected and the school science has been concerned almost exclusively with the content—the body of information. Science is not just content. Science is content plus science processes. If we want to advance in science like other developed countries, science processes should be given due importance like science content in school science curriculum, and students should be encouraged to become personally involved in solving problems and in discovering some science for themselves.

QUESTIONS

1. What was the position of science education in ancient India?
2. "During the British Empire introduction of modern science in India was extremely slow." Discuss.
3. Discuss very briefly how the development of science was greatly accelerated after independence.
4. Discuss the role of the curriculum developers when science is developing so fast in our country.
5. "Knowledge of science becomes double every decade." What should be the role of our science teachers in this context?

3

Status of the Subject

In many countries of the world, the primary school curriculum bears little relation to that of fifty years or so ago. Then the subjects were reading, writing and arithmetic. Now the curriculum is achieved much more as a whole. The primary school curriculum has to a considerable extent become integrated and a large number of good primary school teachers possess a broad background, which enables them to guide their pupils' learning on a variety of topics as often based as the surroundings of the school.

However secondary school curriculum generally consists of a number of separate subjects having little or no coordination between them. This may largely be due to the training received by secondary school teachers and to the public examination system which a strongly subject bounded. An attempt has been made in recent years to bring about an integrated curriculum which has helped to bring various science subjects closer but no effort has been made to consider other areas such as languages, mathematics and social sciences.

Significance of Relationship

No subject can be taught in isolation and so is the case with teaching of chemistry. For an effective learning full advantage must be taken of various correlations and applications of chemistry. In addition to correlation of chemistry with other school subjects and daily life, a lot of correlation is possible with other science subjects. Artificial division of science into various branches is a matter of convenience and not of necessity. Based upon this premise, many educators advocate the implementation of curricula based upon the correlation between various subjects. These kinds of curriculum give more meaning to our class room instructions. Various inventions in chemistry have contributed a lot to the social and physical advancement of our society. Chemistry has contributed a lot to development of some other subjects. In the following pages we will take up the correlation of chemistry with other subjects.

Relation with Language

Chemistry in closely related to language in which it is taught. This correlation arises because of the fact that language provides not simply a way of communicating with others but it is also the vehicle of thought. A student can not grapple with a scientific problem without the use of words. This brings about the importance of encouraging the student to master a language both in its spoken and written forms.

Practical work in science provides a very good opportunity for development of a language. It can be developed by discussion between the student and the teacher and also by discussion amongst students themselves. The written form of the language can be developed by encouraging the students to make their own record of practical work, may be in the form of a diary, instead of copying it from the book or black board.

Relation with Mathematics

Mathematics and chemistry are closely related to each other. Actually speaking mathematics is considered as the mother of all sciences. A thorough knowledge of some fundamentals of mathematics is very useful in understanding certain concepts of chemistry. A closer coordination between the chemistry teacher and mathematics teacher makes the job of teaching chemistry easier.

In physical chemistry such topics as thermodynamics, chemical kinetics, radioactivity etc. Can only be properly understood by using certain mathematical equations. For derivation of such equations the students must be familiar with various sign used for representing certain mathematical operations.

Thus we conclude that there is a close relationship and so there is a correlation between chemistry and mathematics.

Relation with Social Sciences

Chemistry is a highly useful subject for the present day society. Many an inventions in chemistry have a lot of social implications and influences the social thinking of individuals. Knowledge of chemistry is quite useful in dispelling superstitions. The contribution of chemistry in development of society is visible in all walks of our life. Many a luxuries which have now become essential for comfortable living owe their origin to knowledge of chemistry. Teacher can refer to such contributions of chemistry while teaching the social sciences. In history reference can be made to various inventions in chemistry which were used to fight or win wars. Geography depend highly on chemistry for some of its aspects. The two subjects geography and chemistry overlap in various areas particularly in areas of study of rocks, atmosphere, hydrosphere, lithosphere, minerals, rain etc. Present day geography is considered as one of the science subjects.

Relation with Physics

Chemistry and physics both are branches of science and they have a large number of common concepts. Many a laws of chemistry can be quite useful for explanation of certain important concepts in physics. The illustration of common topics in chemistry and physics is given by topics such as nuclear physics, thermal physics, atomic physics etc. Many a methods of chemistry are used for carrying out the experiments in physics. This points to a scope of great cooperation between chemistry and physics teachers.

Relation with Biology

The correlation between chemistry and biology is so large that at present we came across such subjects a "biochemistry". There are many a topics in biology which are quite dependent on knowledge of chemistry, *e.g.* biomolecules, working of various human systems such as blood circulation, digestive system etc.

The knowledge of chemistry is helpful in understanding various diseases and in helping to cure/prevent such diseases.

From the above we find a lot of correlation between chemistry and biology. For a better teaching there should be close cooperation between the chemistry teacher and biology teacher.

Relation with Work and Practice

There is a lot of correlation between chemistry and some of the work experience subjects. It is due to this that in many schools chemistry-teachers are assigned the duties which require them to take some work experience subjects. A few such subjects are:

(i) Candle making.

(ii) Itching.

(iii) Engraving.

(iv) Chalk making.

(v) Preparation of shoe polish and nail polish.

(vi) Preparation of soaps and detergents.

(vii) Preparation of antiseptics, cosmetics etc.

Role of Creativity

Chemistry like physics is an experimental science and so it has grown through inventions and discoveries. These require a lot of creativity. It is possible to fulfill the creative urge of students if chemistry is taught to them using the method 'learning by doing'. For this the teacher is expected to impart chemistry instructions in such a way that students are actively involved in all activities and it places a good deal of responsibility on chemistry teachers.

Creativity has been defined in various ways, however there is one thing in common in all these *i.e.* creativity is a process of change, of getting away from main track, of sensing gaps or disturbing missing elements.

A definite correlation has been shown between creativity and intelligence. A high intelligence does not mean high creativity. According to Guilford, creativity represents patterns of primary abilities, patterns which can vary with different spheres of creative ability. It is generally believed that creativity consists of about 120 abilities, the most important of these being sensitivity to problems, fluency of ideas, originality and redefining.

Some of personality characteristics found in creative persons are:

1. Curiosity

2. Ambition
3. Drive
4. Independence of Judgement
5. Self assertion
6. Imagination
7. Initiative
8. Concern for basic problems
9. Openness
10. High ego strength
11. Emotional stability
12. Less talkative
13. Abstract thinking
14. Non-conformist attitude
15. Capability to take risk.

It is possible to foster creativity, through chemistry, in children if we make use of scientific methods in teaching of chemistry. The iinportant of such methods are problem solving method, project method, laboratory method etc.

We can say that most chemists were creative because they relied on method of discovering new knowledge. For fostering creativity in children the chemistry teacher is expected the perform varied roles. He is expected to perform the following roles:

(i) He should give due consideration to questions and ideas of his students.

(ii) He should put provoking questions in class.

(iii) He must be able to recognise originality and should value such an originality.

(iv) He must foster in his students an ability to elaborate a given point.

(v) He should set more questions and problems for experimentation.

(vi) He should help in developing creative ideas.

(vii) He should have guided and planned experiences.

(viii) He should emphasise for research of truth through experimental research.

(ix) He should choose same investigatory projects in chemistry which give his children an opportunity of self-direction.

(x) He should encourage his students to improvise chemistry apparatus and experiments.

4

The Perspective

Science education forms an integral part of our school curriculum upto the secondary level in the process of Universalisation of Science Education. The National Policy on Education (NPE)-1986 has laid considerable emphasis on strengthening the science education in the school education system. The current situation of teaching of school science content is as Environmental Studies at the lower primary level (classes I—V). Integrated approach of teaching science is followed at the upper primary stage (classes VI—VIII) and secondary stage (classes IX—X) by integrating all disciplines of Science in a natural fashion. The NPE-86 envisages extension through every effort of Science Education not only to those who are in the formal school system but also to those who have remained outside the system under non-formal and adult education programmes. The qualitative improvement in science education depends on many vital components. The teacher is considered as a crucial factor in the teaching-learning process, developing positive attitudes in the learners for better achievement and the formulation and implementation of science education programmes.

The teachers have to discard their traditional methods and usual practices in relying entirely on textbooks. The teaching has to be integrated with environment based on real life situations using local experiences, expertise and resources. The classroom territory has to be expanded over the whole environment so that the activities become supplementary to classroom teaching. If such an approach is systematically implemented with mobilisation of needed resources, it is very much likely that there may be an improvement in our science education at the school stage. Several attempts are being made to improve science teaching in our country, both at the formal as well as non-formal sectors of school education. Some of the innovative experiences in science education at the school level include Nehru Science Exhibition, Science Museums and Mobile Science Vans, Vikram Sarabhai Science Centre Ahmedabad, Kerala Shastra Sahitya Parishad and Ekalavya Project etc. In such innovative experiences and science activities, the learning is fitted to the abilities and interests of learners as there exists an opportunity for individuals initiative, independent or collective study and creativity. A large number of such activities are organised in India. A few of the major activities are listed and described briefly here.

Benefits of Exhibition

November 14 is the birth anniversary of Pandit Jawahar Lal Nehru, our first Prime Minister. Children call him Chacha Nehru, as he used to love them much. We celebrate Children's Day on November 14. On this occasion Science Exhibition is also organised by the NCERT. This exhibition is Science Exhibition. To start with, this exhibition used to be organised at New Delhi where Pandit Nehru used to live, when he was our Prime Minister. Now this exhibition is organised in different states. Children from country, from all States and UTs participate in this exhibition. Nehru Exhibition is organised at the national level.

Every year NCERT announces a main theme and sub-themes for Nehru Exhibition. Children make their exhibits on these sub-themes.

Main Theme: Science in our Environment

(i) Agriculture, Horticulture, Farming and Animal Husbandry.

(ii) Conservation of the Environment

(iii) Health

(iv) Energy Conservation and Needs

(v) Astronomy

(vi) Town and Village Planning

(vii) Machines in the Service of Rural Areas

(viii) Teaching Aids for Science and Mathematics

(ix) Innovations.

This information along with the last date of submitting the entry with the dates of exhibition reaches every school of the country, from States and UTs and from States and UTs to all schools. On information children start working on their projects—static or workin investigatory science projects under the guidance and supervision of the teachers. If a school has a Science Club, it becomes quite active after information under the supervision and guidance of the Science Club.

Teacher's Responsibility

Very often in Nehru Science Exhibition we see some static work or some noble experiments demonstrated by the students with lines and graphs. These are projects but not Investigatory Science Projects. Students should be encouraged and motivated to work on some of science projects, and bring them to Nehru Science Exhibition.

A project may be any purposeful activity. It may be a model working, or experiment.

A project which involves investigation, discovery and finding out which was not known to the student before, is an investigatory investigation is much more than the repetition of a standard Experienced student is to decide what experiments are necessary. He may have to design his own apparatus, if that is not available in the laboratory. He has to search for the appropriate principles, laws, formulae, apparatus and data, and originate a solution to a problem. The student has to behave like a scientist.

Working on an investigatory science project is the way a student can learn science by project method which involves some steps of scientific method like Problem, Hypotheses and Experiment.

Every year more and more investigatory science projects are being seen in Nehru Science Exhibition. It is very encouraging. You should also encourage your students to work on investigatory science projects under your guidance and send it to Nehru Science Exhibition through District and State Science Exhibitions and Fairs.

Benefits from Museums

You must have seen science museums. They are very effective and interesting sources of learning science. What experiences do you get from a science museum ?

Objectives of Science Museum. Main objectives in establishing science museums are:

(i) to help young science learners in understanding concepts of science by play way method.

(ii) to provide a glimpse of past as well as an insight into the future.

(iii) to help schools in their class activities by providing them with a number of equipments and specimens which are otherwise difficult for a single school to procure.

(iv) to arrange extension activities such as field trips, lectures, film shows and exhibitions for the students as well as public.

During the last decade or so, a couple of science museums have been set up in the country including one at Delhi, the Natural History Museum. In Delhi's Pragati Maidan, National Science Centre (National Council of Science Museums) has also been set up few years back. It has several units. 'FUN GAMES' and 'ENERGY' units are very interesting for students. Delhi also has two Primary Science Museums:

(i) Municipal Corporation Children Resource Centre (MCCRC) at R.K. Puram, Sector VI.

(ii) New Delhi Municipal Council (NDMC) Science Centre at Lakshmibai Nagar.

There are good science museums in various parts of the country—Bombay, Bangalore, Calcutta. All these science museums are doing very good Job, carrying out various innovative activities, in science for the improvement of science education. Let us discuss some science museums a little bit in detail.

Nehru Science Centre. The Nehru Science Centre (NSC) is established in Bombay by the National Council of Science Museums. The most important and attractive part of the NSC is a 'Science Park' for children. With green surroundings, the Children's Science Park has exhibited on time, motion, energy, power and work. Also, there are models of railway engines, tram cars, aeroplanes, steam lorries, a windmill and a sun dial. There are birds, animals and fish to acquaint children with nature. While children enjoy the Science Park the most, it also helps them to understand 'what' 'why' and 'how' of the queries, questions and problems haunting their minds.

Nehru Science Centre, Bombay is basically multi-disciplinary in character. Collection of antique exhibits of historic value, presentation of the same through permanent and temporary exhibitions on selected themes, extension activities offering multiple avenues of learning, enjoyment and training to the student community as well as the public, taking science to rural areas through mobile science vans, aiming towards interacting mode of presentation of themes are some of the ways in which the NSC operates. The Science Centre also offers a gallery on 'light and sight'. It presents different principles involved in file process of 'seeing and the vision.' A survey is made of vision, its defects, its importance, its complexities and varieties. Nehru Science Centre also organises extension activities such as science extension in rural areas, film video shows, science seminars for schools, films, video cassettes loan service, amateur weather station, amateur radio classes for children, sky observation programme, astronomical camps, popular science lectures, special science film festivals, aeronautic modelling programmes and training camps for under-privileged children.

Visvesvaraya Industrial and Technological Museums. The Visvesvaraya Industrial and Technological Museum is established in Bombay. It organises various activities and programmes such as motive power gallery (science museum), teacher training, hobby centre, student's science seminars, science quiz, science fair, temporary science exhibitions, science demonstration lectures, mobile science exhibitions, film shows, popular science lectures, etc. The museum has also a regional science centre at Gulbarga.

If you have a science museum at the place where you teach, plan a visit to that museum. If your students go to some places, where there are science museums, ask them to visit them with their parents or if you arrange a field trip to any place, where there is a science museum, take your students there, and see how much science they learn—the science which is not there even in their science books.

Mobile Science Vans. Some science museums have mobile science units, museums on wheel. They are usually sent to the places, where there are no science museums. They are not as big as a science museum. They do not have as many science exhibits as a science museum has. But even this is a very effective source of learning science for students at upper primary level.

Natural History Museum, New Delhi, National Science Centre, New Delhi and Nehru Science Centre, Bombay have Mobile Science Vans. You try to find out whether there is some science museum, not very far from the place you are teaching, and whether that science museum has the facility of mobile science vans. You can ask such museums to send that unit to your school. If you are successful to bring it, your students may visit the science museum even in their school, and can learn a lot of science.

Other Organisations

For improving the quality of science learning in non-formal system of education, Vikram Sarabhai Community Centre was established in Ahmedabad in 1963. The Centre is one of the pioneer organisations in the country providing a variety of out-of-school activities in science for students, teachers and community. It has a team of highly skilled staff which acts as a nucleus and catalyst for various programmes undertaken by the Centre.

The Centre conducts research and innovative programmes for improving education and community life. These programmes include studies on science and mathematics, environmental studies, integrated science and science learning improvement programmes through enquiry approach, mathematics laboratory, teacher orientation, designing and development of teaching and learning material packages.

The Centre also organises programmes for rural as well as urban community. These programmes are related to the problems of pollution, health, security, population, communication, settlement and values. The Centre is basically a community centre where people come with their children and learn science where interested teachers and scientists experiment new ideas in teaching and learning. The Centre organises science seminars, film video shows, popular lectures, exhibitions, sky-gazing through a telescope, etc. The Centre has a library, laboratories, science museum, workshop, science playground, mass media and A.V. facilities for the community. The Centre provides facilities in rocketry and electronics hobbies to children. In science playground, the children get a glimpse of science through play toys, colour filter towards musical pipes, sand, pits, water pond and evolution pillar.

The centre has also started some small extension centres to the rural areas. A mobile van equipped with a laboratory and A.V. materials tours different villages. Science club activities are organised in rural areas based on emphasis on the environmental awareness.

The Kerala Shastra Sahitya Parishad (KSSP) is a voluntary organisation. It was established in 1963. KSSP has around 10,000 members comprising scientists, doctors, engineers, social scientists, teachers, students, workers, peasants and technicians. It has 600 units all over Kerala.

Objectives

1. To popularise science amongst society.
2. To generate science literacy amongst people.
3. To increase community involvement for developing scientific temper in the society.
4. To develop rural technology in the field of energy.
5. To organise health camps, classes and audio-visual campaigns on a wide scale.

Activities. The society has about eight major areas of activities —

(a) Publications, (b) Non-formal Education, (c) Formal Education, (d) Environmental Bridge, (e) Research and Development Wing, (f) Rural Science Forums, (g) Health Brigade, (h) Art and Science.

The details of these activities are as follows.

Publications. KSSP prints a variety of scientific periodicals and books meant for popularisation of science and generation of science literacy amongst the people. These include:

(i) Eureka—monthly magazine for primary classes.

(ii) Sastrakeralam—monthly magazine for secondary School children.

(iii) Sastragathy—monthly magazine for adults.

(iv) Parishad Vartha—monthly bulletin for members.

Non-formal Education. These activities cover a wide spectrum, the main ones being 'Science Campaign' and 'Science Centre.'

Formal Education. KSSP promotes a number of activities aimed at improving science clubs, talent tests and promotion of awareness about the education system amongst the public. The talent tests are:

(i) Eureka Talent Tests—at elementary level.

(ii) Sastrakeralam Quiz—at high school level.

(iii) Sastragathi Talent Tests—at college level.

Environmental Bridge. KSSP was involved m Silent Valley Campaign, Social Forestry Programmes and campaigns against industrial pollution.

Research and Development Wing: Its responsibility is to develop appropriate rural technology in the field of energy, environment etc. A high efficiency *Chulha* (Stove), developed by them has been widely propagated.

Rural Science Forums: KSSP has initiated these forums to prompt villagers to think on their own about their problems and solutions.

Health Brigade: KSSP organises health camps, classes and audio-visual campaigns on a wide scale.

Art and Science: It organises Sastra Kala Jatha and Bharat Kala Jatha.

Ekalavya Science Teaching Project (ESTP) started in 1972 in sixteen rural middle schools of district Hoshangabad in Madhya Pradesh for teaching science through environment based discovery approach. This project was started by Kishore Bharati, in collaboration with Friends Rural Centre, Rasulia with the support of the Department of Education, Government of Madhya Pradesh. A large number of teachers and scientists from various institutions and organisations such as the All India Science Teachers Association, Physics Study Group; Bombay Municipal Corporation; Gandhi Vidyapeeth, Vedehi, Surat District; Lok Bharti, in Gujarat; The Space Application Centre, Ahmedabad; Universities of Delhi, Rajasthan and Indore; The Tata Institute of Fundamental Research, Bombay; Indian Institute of Technology, Kanpur; NCERT, DAV College of Education, Abohar (Punjab) etc. participated in the development of curriculum, workbooks, science kit, other materials and training of teachers.

In 1978 this programme of science teaching was extended to all the 206 middle schools of District Hoshangabad.

Objectives

1. Implementation of introducing innovations as envisaged in Ekalavya Project within the given framework of the Government school system.

2. Encouraging science teaching through discovery approach in Indian schools.

3. Providing science education experiences through environment.

4. Developing ability among students for applying scientific methods in different situations.

5. Developing scientific attitude (scientific temper) among the students.

Curriculum. Keeping in view the objectives of this Ekalavya experiences the curriculum of science teaching has been on process approach rather than product approach. The process approach of learning science provides numerous opportunities to children to explore scientific phenomena of their local environment. Most of the curricular contents have been taken from their environment. Advanced scientific concepts, such as abstract chemical symbols, theoretical concepts of atomic and molecular structure, and human anatomy etc. have not been included in the curriculum because these concepts are beyond the students direct interaction with the environment.

The selection of curricular content is dependent upon:

(a) relatedness to environment, (b) relatedness to the needs, interests and mental level of the students, and (c) possibility of the application of discovery approach.

Through the above mentioned procedure the curricular contents for classes sixth, seventh and eighth was developed. Some of the examples of the curricular contents for various classes are given below:

Class VI	Kuchh Khel Khilwar
	Samuh Banana Sikho
	Hamari Phaslen aur Samuhikaran
	Vidyut
	Ganak Ke Khel
	Jeev Jagad me Vividhata
	Mini, Pathar aur Chattane
Class VII	Ek Majedar Khel
	Jar aur Patti
	Keeron Ki Duniya
	Phaslon Ke Dushman
	Apni Haddi Pahachano
	Aakash Ki Or
	Taraju Ka Sidhant
Class VIII	Jantuon Ka Jivan Chakkar
	Phool aur Phal
	Paudho me Prajanan
	Vargikaran Ke Niyam
	Jantuon ka Vargikaran
	Gasen

Work Book and Science Kit Materials. In this programme, the workbook is introduced in place of textbooks, which is process based. Principles of science are discovered through experiments. Science Kit is very conducive for discovery approach to science teaching.

Teaching Method. Discovery approach of science teaching is the main teaching method followed in this programme. Students learn science through inquiry approach especially by experimentation, discussion and field trips. The whole class is divided into subgroups of four students each known as a Toli. This Toli pattern is also followed in their teacher training programmes. The students perform experiments in their respective tolies, collect and analyse data and draw conclusions on the basis of the guidelines given in the workbook.

Examination. In this Eklavaya Experience the examination is not based on rote memory or recall etc. Independent observation, data collection, data analysis and drawing conclusions have been given due weightage. It also seeks to test the extent of a pupil's readiness to innovate through physical experimentation.

The examination is conducted to test three basic elements of science teaching, namely, scientific skills, scientific attitude (scientific temper) and understanding of scientific concepts and principles.

QUESTIONS

1. When is the Nehru Science Exhibition organised? Who organises it? Where is it organised? Who participate in it? What is the nature of exhibits displayed in it?

2. What is the difference between a 'science museum,' and a 'mobile science unit'? How will you use them in science teaching?

3. Name the science museums you have seen. Where are they located? What did you see in them?

4. Name the city and State where "Vikram Sarabhai Community Science Centre" is situated. What are its activities?

5. What do you mean by "KSSP"? When was it established? What are its activities?

6. What do you mean by "ESTP"? When and where was it started? What are its objectives? What materials were developed under this project?

5

Goals and Purposes

In order to accomplish the task of teaching chemistry, it is essential for us to be very clear about the purpose of teaching chemistry. If we have a clear idea of what is to be achieved them it would be easier to implement any prescribed course in chemistry. This clarity of purpose would also be quite helpful in measuring the effectiveness of teaching chemistry. The purpose of teaching chemistry is based on certain aims and objectives to be achieved. Teacher may use different methods of teaching to achieve the purpose. Many educational reform committees have emphasised spelling out aims and objectives of teaching a particular course of study.

The aim of teaching chemistry refers to the advantages that can be drawn or purposes that can be served by the study of chemistry. The important aims of teaching chemistry are as follows:

Knowledge Aim : The teaching of chemistry should increase the knowledge of the individual and such an increase in knowledge should help him in understanding himself and his environment. Thus this knowledge should help him in his daily life.

Practical Aim : The knowledge gained should be of practical use to an individual. The individual should not only know the principles, and facts but should be able to use these principles in understanding his environment. For this the knowledge should be related to the materials, with which the pupil is familiar and should not be based on obsolete devices and ideas.

Development of Scientific Attitude : Chemistry being a physical science it aims at the development of scientific attitude among the learners. It should be helpful in removing the superstitions, false beliefs, wrong notions spread in the society and cultivate the habits of proper reasoning, observation and experiment action. One of the major aims of chemistry like any other science subject is to develop scientific attitude and science related values amongst students. It should train the student in the method of science and should help develop in scientific temper.

Cultural Aim : Present day civilisation owes much to the development of chemistry and for any further development we have to strive for progressive improvement in the study of chemistry. For this the chemistry be taught in schools in such a way as (i) to grasp the progress in the field of chemistry (ii) apply it for enhancement of our cultural heritage and development of civilisation and (iii) appreciate the study of chemistry in the progress and development of culture and civilisation.

Social Aim : The study of chemistry should help inculcate social virtues among the students for leading a well adjusted social life and contributing significantly towards welfare and progress of society. It should imbibe in him essential social qualities and virtues for becoming a responsible useful citizen.

Vocational Aim : The knowledge of chemistry in the present day world is essential for almost all the professions, and vocations.

To achieve the vocational aim we must prepare our students for the different occupations and vocational courses. This knowledge should also provide them proper opportunity for adoption of some chemistry hobby and engage themselves in small scale industries and self employment projects.

Utilisation of Leisure Time : The knowledge of chemistry should be useful to an individual to learn ways and means of utilizing his leisure hours more fruitfully.

Psychological Aim : Teaching of chemistry provides to an individual various opportunities for satisfying his varying psychological needs and this help him grow and develop as a well balanced individual.

Skill Aim : Like any other science subject, the teaching of chemistry should aim to develop useful skills pertaining to scientific observation, experimentation and practical use of scientific facts and principles.

Selective Objectives

Thurber and Collette have proposed the following criteria for selection of aims.

(i) *Usefulness.* The knowledge gained should be useful to the students in their lives.

(ii) *Timliness.* The knowledge given should be concerned with materials/objects with which student is familiar.

(iii) *Fitness.* The knowledge should fit into sequence that leads him to broad objectives.

(iv) *Appropriateness.* The learning should be appropriate for maturity and background of the students.

(v) *Practicability.* It means that experience required for development of learning should be possible.

The chemistry team of the institute for the Promotion of Teaching Science and Technology (IPST) in Thailand formulated the following broad aims which they felt valid for any science course.

1. To develop an understanding of the basic principles and theories of science.

2. To develop an understanding of the nature, scope and limitations of science.

3. To develop a scientific attitude.

4. To develop skills important for scientific investigation.

5. To develop an understanding of the consequences of science on man and his physical and biological environment.

Aims of chemistry curriculum should be as follows:

(i) To make students interested in chemistry.

(ii) To familiarise the students with the important role played by chemistry in their life.

(iii) To develop in students a scientific culture.

(iv) To provide a training to students in methods of science.

(v) To emphasise upon students the role of chemistry on social behaviour.

(vi) To prepare students for those vocations which require a sound knowledge of chemistry.

(vii) To increase students understanding to such a level that he can understand various concepts and theories which unify various branches of chemistry.

The Distinction

Though the two terms 'aims' and 'objectives' are considered as synonyms and used interchangeably yet in a deep sense there is a lot of difference between 'aim' and 'objective'.

Values and aims are quite inter related and interdependent. We aim at a thing because we value it. The values or advantages that we can draw by achieving a thing become our purposes or aims. These may be taken as the broader purposes or goals or targets that can be anticipated through the teaching of chemistry.

To achieve these aims we have to proceed systematically. For achieving these aims, these are usually divided into some definite, functional and workable units named as objectives. Objectives are, therefore, the ways and means of achieving the aims in a more practical and definite way.

Objectives are the specific and precise behavioural outcome of teaching a particular topic in chemistry. The objective of a topic in chemistry help in realising some general aim of teaching chemistry. The characteristics of a good objective are as under:

(i) It should be specific and precise.

(ii) It should be attainable.

Probably the most common educational objective in the *acquisition of knowledge*. By knowledge, we mean that the student can give evidence that he remembers either by recalling or by recognising, some idea or phenomenon which he has had experience in the educational process. Knowledge may involve more complex processes of relating and judging.

Another important objective is development of *intellectual abilities and skills*. This has been labelled as 'critical thinking' by some, 'problem solving' by others.

Arts or skills + knowledge = ability 'Arts and skills' refer to modes of operation and generalised technique for dealing with problem. The arts and skill emphasise the mental processes of organising and recognising material to achieve a particular purpose. *Intellectual abilities* refer to situations in which the individual in expected to bring specific technical information to bear on a new problem

Objectives are the specific and precise behavioural outcomes of teaching a particular topic or lesson of chemistry. Objectives actually control other factors of chemistry teaching to a great extent, therefore more emphasis be laid on writing objectives in behavioural terms for each unit of class room instructions in chemistry.

"Classification especially of animals and plants according to their natural relationships".

Taxonomy of educational objectives is intended to provide for classification of the goals of our educational system. It is expected to help in the discussion of curricular and evaluation problems with greater precision. It is expected to facilitate the exchange of information about curricular developments and evaluation devices.

Bloom's taxonomy is a classification of instructional objectives in a hierarchy. It is found quite useful in communicating the objectives of a chemistry lesson as also a criterion for evaluation of chemistry teaching. Under this scheme the specific objectives are classified as falling into the following three domains.

1. Cognitive domain objectives.

2. Affective domain objectives.

3. Psychomotor domain objectives.

Area of Influence

Probably the most common educational objective is acquisition of knowledge. Knowledge, as defined here, involves the recall of specifics and universals, the recall of methods and processes or the recall of a pattern, structure or setting.

The cognitive domain can be summarised as under:

Classes	*Instructional Coverage*
Knowledge	Recall and recognition of facts information, principles, laws and theories of chemistry.
(i) Knowledge of specifics	The recall of specific and isolable bits of information.
(ii) Knowledge of terminology	Knowledge of refrents for specific symbols (verbal and non-verbal) *e.g.* to define technical terms.
(iii) Knowledge of specific facts	Knowledge of dates, events persons, places, etc.
(iv) Knowledge of ways and means of dealing and with specifics	Knowledge of the ways of organising, studying, judging, critisising.
(v) Knowledge of conventions	Familiarity with the forms and conventions of scientific papers.
(vi) Knowledge of trends and sequences	Knowledge of the process directions and movements of phenomenon with respect to time.
(vii) Knowledge of Classification and categories	To recognise the are encompassed by various kinds of problems and arguments.

Contd.

Classes	*Instructional Coverage*
(viii) Knowledge of criteria	Knowledge of a criteria by which facts, principles opinions and conduct are tested or judged.
(ix) Knowledge of methodology of evaluation.	Knowledge of scientific methods
(x) Knowledge of principles and generalisations	Knowledge of important principles.
Comprehension	It represents the lowest level of understanding.
(i) Translation	The ability to understand non-literal statements.
(ii) Interpretation	The ability to grasp the thought of the work as a whole at any desired level of generality.
(iii) Extrapolation	The ability to deal with the conclusions of a work in terms of the immediate inference made from the explicit statements.
Application	Application to phenomenon discussed in one paper of the scientific terms or concepts used in other papers.
Analysis	The breakdown of a communication into its constituent elements or parts such that the relative hierarchy of ideas in made clear and/or the relation between the ideas expressed are made explicit.
(i) Analysis of elements	The ability to recognise unstated assumptions, skills in distinguishing facts from hypothesis.

Contd.

Classes	***Instructional Coverage***
(ii) Analysis of relationships	Ability to check the consistency of hypothesis with given information and assumptions.
(iii) Analysis of organisational principles	The organisation, systematic arrangement and structure which hold the communication together.
Synthesis	The putting together of elements and parts so as to form a whole.
(i) Production of a unique communication	Skill in writing, using an excellent organisation of ideas and statements. Ability to tell a personal experience effectively.
(ii) Production of a Plan	Ability to propose ways of testing hypothesis.
(iii) Derivation of set of abstract reiations	Ability to formulate appropriate hypothesis based upon an analy-sis of factors involved and to modify such hypothesis on the basis of new factors and considerations.
Evaluation	Judgement about the value of material and methods for given purposes.
(i) Judgement in terms of internal evidence	The ability to indicate logical fallacies in arguments.
(ii) Judgements in terms of external evidence	Judging by external standards, criteria ability to compare a work with the highest known standard in its field.
(iii) Psychomotor	Development of skills such as of handling pieces of apparatus, their assemblies, drawing diagrams and circuits, repair of apparatus and appliances.

The Knowledge

To impart knowledge in the basic aim of education and so it naturally is the basic aim of teaching of any subject including science. By imparting knowledge of science to the student it is expected that he acquires the knowledge of:

(i) Natural phenomenon.

(ii) Scientific terminology.

(iii) Scientific concepts and formula.

(iv) Modern inventions of science.

(v) Importance of animal life and plant life to man.

(vi) Manipulation of nature by man.

(vii) Correlation and inter-dependence of various branches of science.

(viii) Environment.

Knowledge objective is considered to have been achieved if the student is able to recall and recognise terms, facts, symbols, concepts etc.

The Understanding

This objective considered to have been achieved if the student is able to:

(i) Interpret charts, graphs, data, concept etc., correctly.

(ii) Illustrate scientific terms, concepts, facts, phenomenon's.

(iii) Explain facts, concepts, principles etc.

(iv) Discriminate between different facts, concepts etc. that are closely related to each other.

(v) Identify relationships between various facts, concepts, phenomenon etc.

(vi) Change tables, symbols, terms etc. from any given form to some other desired form.

(vii) Find faults, if any, in statements concepts etc.

The Applications

This objective seems to be the most neglected one in our educational system. The common observation that supports it is that a science graduate fails to insert even a fuse wire in the electric circuit of his house. This objective is considered to have been achieved to a great extent if the pupil can:

(i) Analyse a given data.

(ii) Explain giving reasons various scientific phenomenon.

(iii) Formulate hypothesis from his observations.

(iv) Confirm or reject a hypothesis.

(v) Correctly infer the observed facts.

(vi) Find cause and effect relationship.

(vii) Give new illustrations

(viii) Predict new happenings.

(ix) Find relationships that exist between various facts, concepts, phenomenon learnt by him.

The Skills

This objective can be considered to have been achieved if a pupil learns (i) handling piece of apparatus, (ii) assembling pieces of apparatus for experiment (iii) drawing diagrams and illustrations, (iv) constructing things, and (v) carrying our repairs of apparatus and appliances.

Thus here we aim to develop three types of skill in the pupil. These are (a) drawing skill (b) manipulative skill and (c) observational and recording skill.

The drawing skill is considered to have been achieved if pupil is able to draw labelled sketches and diagrams quickly.

The manipulative skill is considered to have been achieveu if pupil is able to

(i) Keep and handle the apparatus properly,

(ii) Improvise models and experiments,

(iii) Observe various precautions while handling apparatus and doing experiments.

The observational and recording skill is considered to have been achieved if the pupil can

(i) Read correctly the instrument or apparatus,

(ii) Record observations faithfully,

(iii) Make calculations correctly and

(iv) Draw inferences correctly.

The Interests

To achieve this objective the pupil is provided with scientific hobbies and other leisure time activities. By providing such activities our aim is to inculcate, among pupils, a living and substaining interest in environment in which he lives.

The aim is considered to have been achieved if the pupil becomes curious and develops such an interest in science that he is always eager and is on look out to:

(i) Take to some interesting scientific hobby.

(ii) Visit places of scientific interest.

(iii) Undertake some chemistry projects.

(iv) Meet and interact some reputed person in the field of chemistry.

(v) Read scientific literature.

(vi) Collect scientific photographs, scientific biographies etc.

(vii) Participate in science fair, science exhibition, science club etc.

(viii) Actively participate in scientific debates, declamation contents, quiz etc.

The Attitudes

Development of scientific attitude is one of the major objectives of teaching chemistry. The development of scientific attitude makes pupil open minded, helps him to make critical observations, develops in time intellectual honesty, curiosity, unbiased and impartial thinking etc.

This objective is considered to have been achieved if a pupil

(i) becomes free of superstitions and prejudices.

(ii) depends for his judgement only on verified facts and not on opinion.

(iii) is readily willing to reconsider his own judgement when some more facts are brought to his notice.

(iv) has an objective approach.

(v) is honest in recording and collecting scientific data.

The Abilities

By the teaching of chemistry we expect to develop the following abilities in the pupil.

(i) Ability to use scientific method.

(ii) Ability to use problem solving method.

(iii) Ability to process information.

(iv) Ability to report things in a technical language.

(v) Ability to collect scientific data from suitable source and to interpret it correctly.

(vi) Ability to organise science fair, science exhibition, science club etc.

The Appreciation

To achieve this objective the teaching of chemistry has to be done in an evolutionary way. For this the curriculum should include such topics where it is possible to reveal stirring biographical anecdotes, some scientific stories having some incidents of adventure, charm and romance. It is possible to achieve this objective by teaching history of chemistry including life stories of some chemists. This objectives can also be achieved by telling the impact of modern chemistry on life.

This objective of teaching chemistry may be considered to have been achieved if the pupil:

(i) Appreciates the contribution of various chemists to human progress.

(ii) Appreciates the history of development of chemistry.

(iii) Realises the importance of chemistry in modern civilisation.

(iv) Takes pleasure in understanding the progress made by science.

Providing Vocational Career

In the modern world majority of Career Courses depend to a large extent on the basic knowledge of chemistry. Some Vocational Courses can be taken up only by students of science. *e.g.* Engineering, medicines, Agriculture etc. For various courses offered by I.T.I's the knowledge of chemistry is the basic requirement. Thus chemistry opens a vast field of opportunities for taking up any vocational course and choose a career. Not only this the knowledge of chemistry develops in a pupil the manipulative skills and he can easily improvise apparatus and experiments and can use his knowledge and skill to make many a common things like ink, soap, candle, chalk, cosmetics, boot polish, nail polish etc. All these provide the pupil with a profitable leisure time work.

Question of Attitude

One of the major aims of teaching chemistry is the development of *scientific attitude* in the pupil. Following are some of the various aspects included in the scientific attitude:

(i) Making pupils open minded.

(ii) Helping pupils open minded.

(iii) Developing intellectual honesty among pupils.

(iv) Developing curiosity among pupils.

(v) Developing unbiased and impartial thinking.

(vi) Developing reflective thinking.

NSSE (National Society of the Study of Education) has defined scientific attitudes "open mindedness, a desire for accurate knowledge, confidence in procedures for seeking knowledge and the expectation that the solution of the problem will come through the use of verified knowledge".

The views regarding scientific attitude expressed at a work shop conducted by the National Council of Educational Research and Training (NCERT) at Chandigarh in 1971 can be summarised as follows. A pupil who has developed scientific attitude:

(i) is clear and precise in his activities and makes clear and precise statements.

(ii) always bases his judgement on verified facts and not on opinion.

(iii) prefers to suspend his judgement if sufficient data is not available.

(iv) is objective in his approach and behaviour.

(v) is free from superstitions.

(vi) is honest and truthful in recording and collecting scientific data.

(vii) after finishing his work takes care to arrange the apparatus, equipments etc. at their proper places.

(viii) shows a favourable reactions towards efforts of using science for human welfare.

The term *'scientific attitude's* by developing scientific attitude in a person certain mind-sets are created in a particular direction. Such mind-sets may be developed either by direct teaching in schools or by out of school experiences gained by the pupil. Though out of school experiences contribute to a large extent yet according to *Curtis* direct teaching does modify the attitude of young pupil.

Tyler also made some suggestions for planning learning experiences in order to inculcate scientific attitude in the pupil. These are summarised below:

(i) The increase in the degree of consistency of the environment helps in developing and inculcating scientific attitude in the pupil.

(ii) The scientific attitude can be inculcated in a pupil by providing him more opportunities for making satisfying adjustments to attitude situations.

(iii) The scientific attitude can also be developed in the pupil by providing him opportunity for the analysis of problem or situation so that a pupil may understand and then rest intellectually in desirable attitude.

Teacher's Responsibility

The major role can be played by the chemistry teacher in developing scientific attitudes among his students and this he can do by manipulating various situations that infuse among the pupils certain characteristics of scientific attitudes. He can also help in developing a scientific attitude among his students if he possesses and practices various elements of these attitudes. The practical examples given by the teacher leaves an indelible mark on the personality of his students.

Teacher can use one or more of the ways for developing scientific attitude among his pupils.

Making use of Planned Exercises: A large number of exercises for development of certain scientific attitudes are reported by various journals and magazines. Teacher can frequently use such exercises for developing certain scientific attitudes among the pupils. He can also make use of cuttings from newspapers and science magazines and can display such materials on bulletin board so that it is used again and again for direct teaching.

Exercises which are always included in good text books can also be used by the teacher for developing scientific attitude among his pupils.

Wide Reading: On the basis of a study conducted by him, *Curtis* reported, that those pupil who engage themselves in wide reading in science, develop scientific attitudes more than those who study only one textbook. Thus a teacher should encourage his students to read library books and supplementary books on chemistry. For this it is essential that each school at least has a science corner in its library. The teacher himself must be in habit of making proper use of science library so that his students get encouragement for use of science library. The teacher himself be familiar with the latest new titles in his subject and be willing to share his joys of new readings with his pupils. He should refer some suitable books to his students.

Writing about teachers, Rabinder Nath Tagore has observed, "A teacher can never truly teach unless he is still learning himself. A lamp can never light another unless it continues to burn its own flame. The teacher who has come to the end of his subject, who has no living traffic with his knowledge, but merely repeats his lessons to his students, can only load their minds. He cannot quicken them".

Proper use of Practicals Period: A student of chemistry gets many an opportunities for learning scientific attitudes during his practical periods. It is for the teacher to properly use such opportunities for developing scientific attitudes amongst his pupils. Teacher should take extra care to state the problem of the experiment and should present hypotheses on solution. He should practice the proper method of testing the hypotheses. He should actively participate in discussion and interpretation of results after the experiment. He must inculcate in his students the habit to postpone judgements in the absence of sufficient evidence to support a hypotheses.

Personal Example of the Teacher: Personal example of the teacher is perhaps the single greatest force that is helpful in inculcating the scientific attitudes amongst his pupils. Psychologist have found a great tendency amongst the students to copy their teachers. In this regard some have stated, "As is the teacher, so is the student". It is therefore essential that chemistry teacher is free from bias and prejudices while dealing with his pupils. He should have an open mind and be critical in thought and action in his everyday dealings. He should be totally free from superstitions and unfounded beliefs and should be objective and impartial in his approach to his everyday problems. He should be truthful and should have faith in cause and effect relationship.

Study of Superstitions: There are different types of superstitions that still prevail in Indian society. Simply taking of these superstitions and calling them bad and out of date, will not leave a lasting impression on the minds of the pupils. It will be more useful if the teacher can encourage at least a few of his students to carry out practicals on some popular superstitions such as that the presence of a broken mirror in any home leads to disharmony in that home or that if a cat crosses your way when you are going out for some work, then your work will not be done on that day etc. etc.

Such beliefs can easily be discarded by a student if he keeps a broken mirror at his home and finds to his satisfaction that it has not created any type of disharmony in his home. Similarly, other superstitions and misbeliefs can be tested and easily discarded by a student of chemistry. Various researches carried out in the field have drawn the same conclusion *i.e,* by practical survey and study of such common beliefs, students have developed permanent mind-sets or attitudes towards such superstitions.

Co-curriculum Activities in Chemistry: Various co-curricular activities such as organising science club, hobbies club, chemistry society, organising scientific tours and excursions etc. can be taken up by chemistry teacher. Such activities should be properly organised by chemistry teacher under his direct supervision but students be given enough freedom to plan their activities. It will help inculcate in students some desirable scientific attitudes. Co-curricular activities may include making of chemistry charts and models, making of improvised chemistry apparatus etc.

Atmosphere of the Class: A proper atmosphere in the class room provided a desirable atmosphere for inculcating of certain scientific attitudes in the pupils. By a proper class atmosphere we mean that the room is properly arranged and suitably decorated in such a manner that it provides for incentive to the pupil to inculcate the habit of cleanliness and orderliness. In addition to such a congenial physical atmosphere of the class room, the teacher's behaviour also contributes to the development of proper class room atmosphere. For inculcating the scientific attitudes amongst his pupils teacher should encourage them in their various activities. He should also take care to see that his lessons content are such as to encourage the students to ask a large number of intelligent question. He should feel pleasure in answering and explaining such questions and must not snub his pupils for asking so many questions.

Methods of Learning

It has already been pointed out that two basic aims of teaching chemistry are (i) development of scientific attitude and (ii) training in scientific methods.

In previous section we have discussed some ways for developing scientific attitude and in this section our aim is to concentrate mainly on training in scientific methods.

A 'scientific method' is 'a method which is used for solving a problem scientifically. It is also referred as 'the method of science' on or 'the method of a scientist. Sometimes it is called as 'problem solving method'. So far it has not been possible to arrive at any commonly agreed definition of scientific method.

The scientific method of teaching chemistry is based upon the process of finding out results by attacking a problem in definite steps, therefore there cannot be any one 'particular method' but such methods have certain common characteristics.

According to Fitzpatrick, "Science is a cumulative and endless series of empirical observations which result in the formation of concepts and theories, with both concepts and theories being subject to modification in the light for their empirical observation. Science is both the study of knowledge and the process of acquiring and refining knowledge". From this it becomes quite clear that students of chemistry be exposed to the scientific method of finding out.

Scientific method helps to develop in a student the power of reasoning, critical thinking and application of scientific knowledge. It also helps in developing positive attitudes amongst the pupils. A list of such traits as given by Woodburn and Oburn is as under:

(i) A scientist must have an unsatiable curiosity, inquisitiveness and a spirit of adventure.

(ii) He should be capable of independent thinking and be ready to abandon the disproved.

(iii) He should be knowledgeable, enlightened and informed.

(iv) He should possess a power of sound judgement and prudent foresight.

(v) He should possess a high degree of perseverance.

Steps of Scientific Method : Since we don't have any single well defined scientific method so we cannot have any well defined fixed steps for a scientific method. However in general the scientific method of teaching chemistry proceeds in the following steps;

(i) Problem in an area of chemistry learning is identified and well stated.

(ii) Relevant data is collected.

(iii) Certain hypothesis are proposed for testing.

(iv) Experiments are set and done to test the proposed hypothesis.

(v) Prediction of other observable phenomenon are deduced from the hypothesis.

(vi) Occurence or non-occurence of predicted phenomenon is observed.

(vii) From observations, the conclusion are drawn to accept, reject or modify the proposed hypothesis.

Thus the scientific method is a sequenced and structured way of finding out the results through experiments. Various steps of scientific method are discussed here.

Statement of the Problem: A student comes across so many things which arouse his curiosity and he has a large number of questions to ask. A good chemistry teacher always encourages his students to ask questions and tries to answer them in a simple and understandable manner. However in answering a particular question the teacher brings to the fore many new problems and it has rightly been said that, "when we double the known, we quadruple the unknown".

Most of the question asked are about 'what?', 'why?' or 'how?' type and these can be conveniently classified as under:

(i) 'what' type of questions are *predictive*

(ii) 'why' type of questions are *explanatory*

(iii) 'how' type of questions are *inventory*

The most important thing in a scientific method is a simple and well defined statement of the problem. The statement of the problem be such that it clearly defines the scope of the problem as also its limitations.

Data Collection: When the problem has been stated in clear terms an effort be made to collect the data from as many different source as is possible. Such data may be available in books in chemistry, library which are an important source for data collection. Data may be collected by use of certain instruments etc. and observations. In data collection an effort be made to minimise the errors that are likely to be caused due to apparatus and instruments used *(mechanical errors)* and those which are likely to be caused due to personal bias *(personal errors)*.

Proposing a Hypothesis: On the basis of collected data a tentative hypothesis is proposed for testing. A hypothesis is in fact a certain tentative solution to the problem. The hypothesis should be proposed only after an objective analysis of the available data because any number of hypothesis can be proposed for a problem. For an objective analysis the student be given a training so that he is free from all his bias towards the problem.

Conducting Experiments: After a hypothesis has been proposed suitable experiments are designed to test the validity of the hypothesis. From the observations of such experiments the validity of the hypothesis is tested. The experiments will show the occurrence or non-occurence of the expected phenomenon and from this we will be able to accept or reject or modify the hypothesis.

The Advantages

Some of the advantages of scientific methods are:

(i) Students learn chemistry by their own experiences and the teacher is just a guide who provides them an opportunity and proper environment for learning chemistry.

(ii) It trains the students to identify and formulate scientific problems.

(iii) It gives enough training to students in techniques of information processing.

(iv) It develops in students the power of logical thinking as he is required to interpret data in a logical way.

(v) It helps to develop an intellectual honesty in the student because he is required to accept or reject the hypothesis on the basis of evidences available.

(vi) It helps the students to learn to see relationships and patterns amongst things and variables.

(vii) It provides the students a training in the methods and skills of discovering new knowledge.

The Disadvantages

Some important disadvantages of scientific methods are as under:

(i) It is a long drawn out and time consuming process.

(ii) It can never be a full fledged method of learning chemistry.

(iii) Majority of chemistry teachers cannot implement it successfully because of their back of exposure to such a method.

(iv) It is not suitable for all students as it suits only bright and creative students.

Summary : We summarise the aims and objectives of teaching chemistry at various stages in India. These are taken from the reports submitted by various commissions, committees etc. which form the basis of chemistry curricula to be taught at various stages of school education.

In 1950 a report was published by Ministry of Education, Govt. of India which listed the aims of teaching of chemistry (science) in schools as follows:

Upto Middle School Level

(a) To develop interest in nature and environment.

(b) To develop creativeness and inventiveness of students.

(c) To inculcate scientific methods.

(d) To develop ability to generalise facts.

(e) To make them understand various social implications of chemistry.

(f) Development of some chemistry-based hobbies and leisure time activities.

At Secondary Level

(a) To enable students to adjust to their environment after understanding it.

(b) To help them to get a feel of scientific methods.

(c) To develop scientific attitude and scientific temper in them.

The major objectives of the syllabus developed, for secondary-schools in India by, National Council of Educational Research and Training (NCERT) are as under:

(i) To strengthen the concept developed at a secondary level and further develop near concepts to provide a sound back ground for higher studies.

(ii) To develop a competence in students to offer professional courses like engineering, medicines etc. as their future career.

(iii) To acquaint the students with different aspects of chemistry used in daily life and enable them to recognise that chemistry plays an important role in the service of man.

(iv) To expose the students to different processes used in industries and their technological applications.

(v) To provide relevant content materials useful for vocational courses.

(vi) To develop an interest in students to study chemistry as a discipline.

6

Teacher's Status

"Teacher! I saw a satellite!" "Look! One baby mouse is white!" "Can we attempt the experiment this way?" "I have collected all the data—use of log tables and even slide rule is too time consuming and out of date for calculations. May I use a calculator? Please! Now we have computers."

Everyday, they are different—but everyday greetings like these begin the day for science teachers. One day it is satellite, one day the moon, one day how the seesaw operates, one day the why of earthquakes—or another question everyone is asking. Science teachers as well as students; government as well as commentators—everyone needs the answers to questions in science.

It is a wonderful world when you look at it this way—and science is like a key that opens doors everywhere—on the bottom of the ocean—with the astronauts in space—with the plants, animals, and man here on earth—with thousands of conveniences of our daily lives. Learning about science—as you already know—

stretches your mind, but gives you a lot in return. You always have something new to bring to your classes. And you as a science teacher have many satisfactions that come from working with young students.

You have been a practising science teacher and now you are ready to be regular science teacher, teaching a class of physics or chemistry, biology or integrated science, you will surely like to be a good science teacher. This is the right time to ask yourself some questions like these:

- What does it take to be a good science teacher?
- How can I use my teacher's training to be a good science teacher?
- Where can I get some more information about good science teaching?
- What things should I do to begin with to be a good science teacher?
- What should I do to be happy as a science teacher?

Role of Teacher

Everyone needs an education in science in order to understand the world and to be a good citizen and good worker. Scientists, leaders in government, industrialists, and the public in general need to be able to understand scientific problems and to make good judgements about them. We need scientific literate and informed people to make decisions about such things as major government projects in space exploration and medical science; health problems, environmental and population problems, such as air and water pollution; diseases such as cancer or tuberculosis and epidemics; or the possible effects of chemical pesticides upon us and our environment.

More and more people will need science in their work, and they must depend on teachers to help them get the knowledge they will need. Each new "breakthrough" or the expanding frontier of science and technology, and the public's need to be informed, require a never-ending supply of well-trained people to fill the ranks. This is also true in engineering, medicine and other fields related to science. The next generation of youngsters, as it passes through school, must be as well-educated in and about science as its talent will permit if we are to continue to prosper in our democratic society. Moreover, the ever-increasing scientific and technological progress challenges man's ingenuity to improve his methods of processing, storing, retrieving and reporting all sorts of scientific information. In a sense, we need "new breeds" of people who can combine science with other talents, as for example, the scientist, librarian and the science reporter.

Good science teaching is one of the most valuable ways to meet this urgent need for science-educated citizens and workers. Enthusiastic, intelligent, and well-educated science teachers inspire and prepare students to investigate the great questions of science and the questions raised by the scientific discoveries which affect us and our society.

Mainly through the inspiration of devoted science teachers, great number of students develop lifelong scientific interests and learn to appreciate and understand the nature of science and its usefulness to mankind.

The science teacher derives great satisfaction from this special importance of his work. And added to it is the great satisfaction that comes from mastering a field of scientific study and from affording a special kind of service to others.

So, when you ask yourself: "What do science teachers do?" You may answer:

"Everyone today needs to understand what science is about. We need it because discoveries in science affect many aspects of society. To be responsible and useful citizens, we must be able to judge how the discoveries of science may be best used for ourselves and society. Many among us will make a career of science, engineering, or technology. We need a good foundation in science to become competent in our chosen work. Therefore, we can say in general that a science teacher, at whatever level he is teaching, is laying the foundations for an understanding of science."

A science teacher is a teacher first and then a science teacher. As a teacher he should be understanding, sympathetic. Teachable and free of prejudices. Science teaching requires a sound knowledge of the subject and a real interest and ability in sharing this knowledge with others. It also requires keeping up-to-date. The work is not easy, but it is creative, and it can be extremely exciting and satisfying. It takes thought, energy, and enthusiasm. Here is an example of what one good school expects of its science teachers:

1. He plans his work for the whole year before school starts its session.

2. He plans his lessons well in advance before he enters the classroom for teaching science.

3. He knows the various practical skills needed in his particular field.

4. He guides and assists his students in their laboratory experiments, field trips and projects. He is also responsible for the safety of the students and the condition of the laboratory and its equipment. He decides in what way each aspect of his subject can best be taught. Some of the ideas and problems can be most effectively presented by demonstration, others through student investigation, field trips, or by films or lectures. He also masters the finer art of asking the questions that will stimulate his students to think for themselves and to search for answers.

5. He organises class procedures, plans the use of the laboratory, discusses with the class, and helps his students learn how to find information outside the classroom in school and city libraries, local industries, or by consulting experts in the community.

6. He encourages his students to develop a lasting interest in science, its changes, and its methods. He guides both individual and group projects and help the students participate in various programmes of awards and recognition. He may sponsor a school science club and with the help of other science teachers and students, organises a science fair or plan field trips.

7. He keeps both himself and his students informed on the most recent developments in his field. He draws his students attention to the social, economic, political, and other aspects of the relationship between society, people and modern science. Thus he and his students always see that science study is not limited to the classroom.

8. He attends orientation courses in his field (seminars, workshops, summer institutes), as well as science conventions and science conferences for his academic growth.

9. He participates in science curriculum development programme conducted by various agencies like DIETs, SCERTs, LASEs and NCERT etc.

10. He conducts action research in science education and actively participates in innovative science programmes for better science teaching.

11. He always efforts for quality science education to his students. Now you can figure out yourself what should be the competencies for a good science teacher (Physics / Chemistry / Biology).

Checking the following questions may help you decide, how good a science teacher you are going to be.

		Yes	***No***
1.	Do you enjoy reading and studying in science as well as in other academic areas?	.....	.....
2.	Have you been a good science teacher?	.....	.....
3.	Do you like working with students and other science teachers?	.....	.....
4.	Do you like to help your students?	.....	.....
5.	Do your students and other science teachers like to work with you?	.....	.....
6.	Could you enjoy and adopt to a career in which the subject is constantly and rapidly changing because of new discoveries?	.....	.....
7.	Do you/ on your own, sometimes do more work than is required by teachers?	.....	
8.	Have you developed science projects and participated in science fairs?	.....	
9.	Do you like to find answers to problems on your own?	.....	
10.	Do you have a sense of humour?	.....	.....
11.	Can you accept criticism and profit from it?	.....	.....
12.	Do you consider that preparing students for living in the World of the future is a challenging job?	.....	.....
13.	Are you developing the ability to read quickly and with comprehension and to express yourself clearly and interestingly in speaking and writing?	.....	.....
14.	Are you in good health, both physically and mentally?	.....	...

If you have checked thoughtfully YES for most of the above questions, you very likely would enjoy teaching science at any level, and you should read on to learn how you prepare for this type of exciting career.

Teacher's Qualities

After completing your teacher training programme, this is the right time to ask yourself whether you possess the basic intellectual and personal qualities necessary to be a good science teacher. Do you have the attributes, such as integrity, drive, and a high sense of responsibility, which are valued in teaching specially science teaching. Besides these, there are special characteristics of successful and happy science teacher.

7

Organised Teaching

Percentage of our science teachers (Physics, Chemistry and Biology) have good background in their respective subjects, and they take interest in teaching and trying out new science programmes. Even in the schools which have enough science equipment and the science labs are well equipped, the lecture-demonstration method is the one most commonly used, as our science teachers are heavily loaded and they get a very little or sometime even no time for planning their lessons during school hours. Our schools have usually 8 periods of 30-40 minutes duration. Regular teachers also work as substitute teachers (where in America if a teacher is on leave, a substitute teacher is invited from outside to teach his classes) during their planning periods besides their teaching 6-7 (out of 8) periods daily, six days a week (where in Western countries schools work five days a week), and Sunday is the only holiday.

Our textbooks are written in a traditional way. If our science teachers could have well sequenced programmed materials, it would be a great help to them, as mentioned above, they are heavily loaded. Also if the programmed materials provided activities of 20-25 minutes duration, students would have more opportunity to do the experiments themselves even in smaller periods. So well sequenced programmed materials would be more useful for teachers to teach and for students to learn than the materials now available.

The Meaning

Let us look at some of the definitions of programmed instruction.

(1) Cronbach describes a programme of instruction as follows: "A programme is a pre-arranged sequence of explanations and questions. A programme whether for a brief unit or for an entire course, is a carefully planned progression of ideas, beginning with elementary notions and working upto relatively complex theories or applications.

(2) According to B.F. Skinner a programmed instruction is simply a matter of breaking the material to be learned into easy steps, arranging steps in logical order with no gaps, making sure the student understands one step before moving to another and then incidently, making sure that he is successful.

(3) According to Espich and Williams a programmed instruction may be defined as a planned sequence of experiences, leading to the proficiency, in terms of stimulus response relationships. By this definition a programme is an educational device that causes a student to progress through a series of experiences, which lead to the students proficiency. The experience

here is student's own experience in the learning process, not just the teacher telling. Planned sequence determines what experiences and in what order should occur. What is the student supposed to be able to do after completing the programme? How well? How quickly? All such questions implied in leading to proficiency in terms of stimulus-response relationships refer to the basic behavioural science concepts on which programmed instruction is based, and which are taken into consideration when a programme is written.

Utility of Psychology

No one knows for certain how or why programmed instruction works, but it is generally agreed that basic behavioural psychology is somehow involved. At least the originators of programmed instruction attributed its success to some basic tenets of behavioural psychology.

The results of the change in behaviour called learning are observable or measurable. All behavioural changes of students as a result of learning may be of three types: (1) Psychomotor; (2) Cognitive; and (3) Affective (Chapter 3). (1) and (2) are comparatively easier to measure than (3).

In any teaching situation for effective learning the best method is:

1. Present the stimulus to the student.
2. Help the student to make the desired response to the stimulus by giving him clues, by leading him towards it, or by telling him the response itself.
3. When the student makes the desired response to the stimulus, immediately reinforces that response.

Programmed instruction takes advantage of the basic human drive for success. The programmer guides the student toward making the correct response. He then shows or tells the student that he has given the desired response—that he has been successful. Each time he makes the correct response, he is positively reinforced by being told that he is correct, his drive for success is satisfied. Each time his drive for success is satisfied, the probability increases that he will make the correct response to the given stimulus in future situations.

The Advantages

The programmed instruction does have a number of advantages over conventional methods. It allows for individual rates of learning and it gives immediate reinforcement of correct responses. According to Burner the technically most interesting features of automatic devices are that they can take some of the load of teaching of the teacher's shoulders. For this time the teacher can be used by his students who need him and whenever they need him. According to James this type of teaching is very frustrating and tiring but extremely rewarding. When science teachers have worked as long at individualized instruction as they have to make grouped instruction workable, the rewards may be proportionally increased.

Positive Aspects

In teaching science, a continuing area of concern to educators has been the problem of integrating laboratory experimentation with instruction in scientific theory. The need to individualize course content for students is also recognized. Programmed instruction may offer one solution to both of these difficulties. If the science curriculum can be programmed so that individual rates could be dealt with more effectively, and if laboratory material can be developed which enable students to conduct experiments effectively on an independent basis, laboratory experi mentation may be integrated more satisfactorily into the typical science course.

A summary of the subject area in which programmed instruction has been used reveals that these areas which require laboratory activities are rare. Cowan indicated that there were no auto-instructional materials available in physics that provided students with laboratory experiences Cowan and Siddiqi developed and used such materials in their research studies.

Hundreds of good science textbooks are available in the market They are not designed to teach but to convey information to the student. In a programme a programmer determines for a student, what he should and what he should not assimilate. In a programme, the student is guided along a path and given those experiences that will cause him to learn those things. No such guidance is given with a textbook Adjunct programming can be a link between programmed instruction and a good textbook. It combines some of the progressive features of programmed instruction with the comprehensiveness of textbooks. The goal of adjunct programming is to enable the student to learn as effectively as possible from a good textbook.

An Adjunct Programme may be one of two types (1) The text itself is kept intact and the programme is supplied as a separate unit, or (2) sections of the textbook are extracted vibration and used in the programme as the basic information. The most popular procedure as used by Cowan and Siddiqi is to leave the textbook intact. Their programmed materials which they developed for the PSSC Physics Text, provide a very good example of adjunct programming. They used these materials in their research studies. Students using the auto instructional materials studied from the same text, worked the same problems, viewed the same films, and performed the same experiments as regular physics students. Cowan did not use a qualified physics teacher in the classroom to direct their studies. The use of auto instructional materials prepared for this study evidently provided a good physics programme, as students using these materials were able to demonstrate as well in achievement as those not using these materials but taught by a qualified PSSC physics teacher. In Siddiqis study students of experimental group used the auto-instructional materials under the guidance of their physics teacher, while, the students of the control group did not use such materials, they were taught by their teacher in a conventional way. The experimental group demonstrated statistically higher mean level of achievement in PSSC physics.

Different Views

Siddiqi developed an Adjunct Programme for PSSC physics. It was student self-directed study guide m PSSC physics. Here are the comments (unedited) of some of the students who used it.

Strengths

1. "Students can do whenever they want to do. If we do not understand anything now, we can do it some other time."
2. "When you learn on your own, you are more apt to remember it."
3. "Student learns more if he goes on his own rate. In conventional classes superior students get frustrated. They want to go ahead. Slow students don't understand. Teacher stops the whole class. With these materials superior students also have an opportunity to go ahead. Every student is able to go on his own rate, not waiting for anybody, not trying to catch up with anybody."

4. "Teacher gets more time, when he can go and help other students. He goes to the student who has problems, and not to the student who does not have any problem; so he does not waste his time. Teacher's time can be put to better use. We do not need the teacher all the time. We only use him when we need him. Teacher can enrich the student more when he talks individually."

5. "PSSC text was easy to understand with these materials. Text was too complicated, and the guide made it simplified."

6. "Guide has more hints to do labs than does the lab manual."

7. "Self-tests were good. When I missed something, I went over it and it helped me very much to learn physics."

Weaknesses

1. "These materials were unable to motivate those who were not self-motivated to use these materials."

2. "This approach did not work for lazy students. Some students need a push and without a teacher always available to keep them working they have a tendency not to study. It's hard for some people to pace. It requires more discipline to work on your own."

3. "This approach is good for those who want to learn by this approach. If the student is the type who cannot go on his own, this approach is not good; he should (probably) be taught by the conventional method."

4. "Sometimes a student is isolated by this approach—text guide, problems, lab, text, guide; that's all he has to work with."

5. "Physics is such a hard subject that you have to have more teacher help."

6. "I did not like it too much, because I am lazy by nature."

Teachers who taught the students using programmed instruction materials also pointed out some strengths and some weaknesses of mis approach of teaching.

Strengths

1. "These materials are excellent for those who have difficulty with PSSC physics."

2. "These materials enabled some students to achieve what they were unable to achieve without them."

3. "These materials developed in students an attitude of doing independent work, which they will need when they will go to College and University."

Weaknesses

1. "Only those students who feel responsibility learn more by this approach."

2. "These materials are not good for slow learners."

3. "Some students go behind and behind not because they were not capable to move but because they were lazy. Some students can read but they don't want to read. There is a problem of keeping the student motivated."

4. "This approach is more time consuming. Some students took more time to finish the same material."

5. "When working on their own only half of the students use their time effectively while others use their time doing something else."

The conclusion of a group of students regarding the auto-instructional material was: "With autoinstructional materials, the text, and a teacher to explain, the self-study would be most practical and best for students."

In general most of the students and teachers liked the programmed instructional materials and this approach of instruction. They pointed out individualisation, independence, self-pacing, self-evaluation, and better use of teacher's time as some of the major strengths of this type of instruction. More than half of the students and teachers liked individualised instruction. They reported that using this approach for instruction, the teacher got more time to interact with individual students. They pointed out that the PSSC Physics Students Guide made PSSC Physics easier for the students. They also pointed out that this approach of instruction was best suited for superior students. More than half of the students liked the idea of moving according to their own pace, but the teachers did not like this idea very much, as it was hard for them to cope with their students, when different students were doing different things in the same period. Most of the students and all the teachers pointed out that lack of self-motivation was the main weakness of this type of instruction.

Various Units

Some of you as a science teacher will like to use programmed instruction materials in your science classes. Usually programmes are not available in the market. Though developing programmed instruction materials is a time consuming job, yet some of you will like to develop some programmed instruction units. It will be a good project for your instruction.

Before developing programmed instruction units, it will be nice if you learn how to develop them. One such programmed unit "Structure of an Atom" is given here as a Sample. You go through it, and after that we will discuss, how it is developed.

Structure of an Atom
(A Programme Unit)

Instructions to Students

This is not a test. This is a Programme designed for you to learn the Structure of Atom.

In this Programme you will find numbered paragraphs.

These paragraphs are called Frames. Read each Frame carefully and answer the questions given after each Frame.

Check your answers from the Answer key given at the end of the programme.

Frame 1. Matter is made up of very small particles called atoms. The smallest particle of matter is........

If answer is correct go to Frame 2, otherwise Repeat Frame 1.

Frame 2. Atom is made of various particles called electrons, protons and neutrons. Electrons are negatively charged particles. Protons are positively charged particles. Neutrons have no charge, that is, they are neutral.

Complete the following sentences:

(a) The positively charged particle of an atom is.....

(b) An electron has ...... charge.

(c) A neutral atomic particle is called.......

If answers are correct go to Frame 3 otherwise Repeat Frame 2.

Frame 3. Check the statement below that is true. A proton is:

(a) positively charged.

(b) negatively charged.

(c) neutral.

(d) sometimes positive, sometimes negative.

If answer is correct go to Frame 4, otherwise Repeat Frames 2 and 3.

Frame 4. Check the statement below that is true. A neutron is:

(a) positively charged.

(b) negatively charged.

(c) neutral.

(d) sometimes positive, sometimes negative.

If answer is correct go to Frame 5, otherwise Repeat Frames 2 and 4.

Frame 5. Check the statement below that is true. An electron is:

(a) positively charged.

(b) negatively charged.

(c) neutral.

(d) sometimes positive, sometimes negative.

If answer is correct go to Frame 6, otherwise Repeat Frames 2 and 5.

Frame 6. The weight of a proton is approximately equal to the weight of a neutron, and the weight of an electron is so small compared to the weight of a proton or a neutron that it can be neglected.

Check the statement below that is true.

The weight of a proton is approximately equal to the weight of:

(a) electron.

(b) neutron.

(c) each of them.

(d) none of them.

If answer is correct go to Frame 7, otherwise Repeat Frame 6.

Frame 7. The charge of an electron is equal and opposite to the charge of a proton. If the charge of an electron is -4 units, then the charge of a proton should be:

(a) –4 units

(b) +4 units

(c) more than 4 units

(d) less than 4 units

If answer is correct go to Frame 8, otherwise Repeat Frame 7.

Frame 8. If the charge of a proton is $+x$ units, the charge of an electron should be.......

If answer is correct go to Frame 9, otherwise repeat Frames 7 and 8.

Frame 9. An atom can be assumed to be spherical in shape. At its centre there is a very small space (compared to the atom as a whole) containing protons and neutrons. This space is called nucleus.

Check the statement below that is true. The nucleus of an atom contains:

(a) protons only.

(b) neutrons only.

(c) protons and neutrons both.

(d) none of them.

If answer is correct go to Frame 10, otherwise Repeat Frame 9.

Frame 10. Around the nucleus some hollow spheres of different sizes can be assumed. The space between any two of such hollow spheres is called a shell (Fig.).

∇—boundary of first hollow sphere.

Δ—boundary of second hollow sphere.

Fig.

Check the statement that is true.

(a) A is a shell.

(b) A is a shell.

(c) Both A and Â are shells.

(d) None of them are shells.

If answer is correct go to Frame 11, otherwise Repeat Frame 10.

Frame 11. As the number of electrons is equal to the number of protons and the charge of an electron is equal and opposite to the charge of a proton the atom as a whole is neutral.

Example—Sodium atom has 11 electrons and 11 protons. If charge of an electron is—x units, the charge of a proton will be $+x$ units.

Charge of 11 electrons = $-11x$ units.

Charge of 11 protons = +11x units.

Therefore total charge of sodium atom is equal to –11x plus +11x = 0.

Complete the following statements:

(a) A neutral atom has 40 electrons, then it should have .... protons.

(b) If an atom has 10 electrons but only 9 protons, then the atom .... neutral (will be, will not be).

If answers are correct go to Frame 12, otherwise Repeat Frame 11.

Frame 12. Nucleus has protons and neutrons. Around the nucleus there exist electrons in various spherical shells.

Complete the following statements:

(a) Electrons exist in the.........

(b) Protons exist in the..........

(c) Neutrons exist in the..........

If answer is correct go to Frame 13, otherwise Repeat Frame 12.

Frame 13. In a particular shell the electrons are assumed to be moving in various circular orbits. The total number of electrons in all the shells of an atom is equal to the number of protons in its nucleus. This number *i.e.* the number of protons or electrons in an atom, is called atomic number.

(a) Sodium has 11 protons in the nucleus. The total number of electrons in all the shells of sodium will be........

(b) The atomic number of sodium is.......

If answers are correct go to Frame 14, otherwise Repeat Frame 13.

Frame 14. The atomic number of oxygen is 8. What will be the number of protons in the nucleus, and the number of electrons in the shells?

(a) Number of protons is......

(b) Number of electrons is.......

If answers are correct go to Frame 15, otherwise Repeat Frames 13 and 14.

Frame 15. The sum of the number of neutrons and protons is equal to the atomic weight of an atom.

Example. Oxygen has 8 protons and 8 neutrons. Thus the atomic weight of oxygen is 8 + 8 = 16.

Sodium atom has 11 protons and 12 neutrons. Check the statement below that is true. Atomic weight of sodium is:

(a) 11

(b) 12

(c) 23

(d) 1

If answer is correct go to Frame 16, otherwise Repeat Frame 15.

Frame 16. Nitrogen atom has 7 protons and 7 neutrons. The atomic weight of nitrogen is..... .

Frame 17. The atomic weight of hydrogen is 1. It has 1 proton, then the number of neutrons in hydrogen atom is......

If answer is correct go to Frame 18, otherwise Repeat Frames 15,16 and 17.

Frame 18. The number of shells in an atom may vary from 1 to 7, that is the number of shells in an atom cannot exceed 7, though it can be less than 7 i.e. (6, 5,4,3, 2,1) depending upon the total number of electrons in a particular atom.

The number of electrons present in each shell is governed by 2 rules. *Rule 1*—The nth (*n* is the number of shell) shell cannot have more than $2n^2$ n electrons.

Example

For the first shell $n = 1$, therefore $2n^2 = 2$. f = 2

It means 1st shell cannot have more than 2 electrons (if the atom has more electrons they will be in the other shells), though it can have less than 2 (if there are not enough electrons in a particular atom).

For the 2nd shell $n = 2$, Therefore $2n^2 = 2 .2^2 = 2 .4 = 8$

It means 2nd shell cannot have more than 8 electrons though it can have less than 8 (as explained above).

According to this rule, there should not be more than:

(a)..... electrons in 3rd shell.

(b)...... electrons in 4th shell.

(c)..... electrons in 5th shell.

(d)..... electrons in 6th shell.

(e)..... electrons in 7th shell.

If answers are correct go to Frame *19*, otherwise Repeat Frame 18.

Frame 19. Oxygen has 8 electrons. According to the above mentioned rule write down the number of electrons in different shells.

Shell 1.......

Shell 2.......

Shell 3.......

If answers are correct go to Frame 20, otherwise Repeat Frames 18 and 19.

Frame 20. The atomic number of chlorine is 17. Write down the number of electrons in different shells.

Shell 1.......

Shell 2.......

Shell 3.......

Shell 4.......

If answers are correct go to Frame 21, otherwise Repeat Frames 18, 19 and 20.

Frame 21. *Rule 2*—Last shell cannot have more than 8 electrons in an atom, and the last but one shell cannot have more than 18 electrons. Complete the following statements:

(a) An atom has 5 shells. The number of electrons in its 5th shell should not exceed......

(b) The number of electrons in its 4th shell should not exceed ....

If answers are correct go to Frame 22, otherwise Repeat Frame 21.

Frame 22. The atomic number of sodium atom is 11 and its atomic weight is 23. Write down the number of the following items of a sodium atom.

(a) Electrons...........

(b) Protons.........

(c) Protons and neutrons........

(d) Neutrons..........

(e) Electrons in 1st shell........

(f) Electrons in 2nd shell..........

(g) Electrons in 3rd shell...........

(h) Electrons in 4th shell...........

If answers are correct go to Frame 23, otherwise Repeat Frames 13-22.

Frame 23. The atomic number of carbon is 6 and its atomic weight is 12. Complete the carbon atom by putting at proper places the number of electrons, protons and neutrons.

If answer is correct congratulations, you learned what we wanted to teach you through this programme. If not please repeat Frames 13-23.

ANSWERS

Frame 1

Atom

Frame 2

(a) proton

(b) negative

(c) neutron

Frame 3

(a) positively charged

Frame 4

(c) neutral

Frame 5

(b) negatively charged

Frame 6

(b) neutron

Frame 7

(b) +4 units

Frame 8

X units

Frame 9

(c) protons and neutrons both

Frame 10

(b) B is a shell

Frame 11

(a) 40 protons

(b) will not be

Frame 12

(a) shells

(b) nucleus

(c) nucleus

Frame 13

(a) 11

(b) 11

Frame 14

(a) 8 protons

(b) 8 electrons

Frame 15

(c) 23

Frame 16

14

Frame 17

0

Frame 18

(a) 18

(b) 32

(c) 50

(d) 72

(e) 98

Frame 19

shell 1=2, shell 2=6, shell 3=0

Frame 20

shell 1=2, shell 2=8,

shell 3=7, shell 4=0

Frame 21

(a) 8

(b) 18

Frame 22

(a) 11

(b) 11

(c) 23

(d) 12

(e) 2

(f) 8

(g) 1

(h) 0

In developing this unit the System Approach was used. The System Approach (Fig.) has the following steps:

1. Identify problems and state terminal objectives.
2. Conduct task analysis.
3. Describe entry behaviours of students.
4. State sub-objectives in behavioural terms.
5. Develop evaluation instruments.
6. Determine instructional sequence.
7. Select appropriate media and instructional procedures.
8. Develop instructional materials.
9. Conduct formative and summative evaluation.

Given the atomic weight and atomic number of an element the students will find out the number of protons and neutrons in the nucleus and the number of electrons in the various shells of the atom.

Hierarchical methodology (Fig.) was chosen for the programme Structure of an Atom. There is an ordered relationship between sub-skills which ultimately help to learn the terminal objective. The simple things are given first that makes the basis for more complex concepts.

Behavioural Aspects

The assumed entry behaviours for this programme were the following:

1. Knowledge of some common elements such as oxygen, hydrogen, nitrogen, carbon, sulphur, sodium, chlorine etc.
2. Simple addition, subtraction, multiplication and division.
3. Knowledge of positive and negative charges.
4. Basic skills of reading, writing, generalizing, and following directions.

Behavioural objectives were prepared in such a way that they were consistent with each task in the task analysis.

All the behavioural objectives contain an observable response, important condition under which the behaviour will be evaluated, and the criterion for satisfactory performance.

Test Devices

The test was made to test all the eleven behavioural objectives. It contained 16 test items. There were one or two test items to test each behavioural objective. All the test items are multiple choice tests and seems to be appropriate to test the task required in the behavioural objectives. The language used in writing test items was kept easy enough to be understood by the students.

This test was used as a Pre-Test, in which the items are sequenced m the same manner as they would appear in the programme (in the Sequence of Tasks and Behavioural Objectives). The same test can also be used as Post-Test (though a separate Post-Test is preferred), whose items should be scrambled, so that they are not in the same order as in the programme. In the multiple choice questions the order of correct responses was random.

The frames were constructed in the same sequence as shown in the task analysis. The frames were sequenced in such a way that by completing all the frames the terminal objective might be achieved by the students.

Medium in Effect

The strategy of linear programming was chosen. The self-learning unit was an instructional programme, having enough practice frames.

Examination Programmes

Formative Evaluation. After writing the programme and the tests (pre-test and post-test) one IX class student was selected for one-on-one evaluation. This student was of an average ability. Two copies of the programme were prepared, one for the student and other for the programmer. When the student was using the programme, the programmer noted down his reactions and suggestions, which seemed to be important in revising the programme. The following feedback was received from one-to-one evaluation.

1. There was smooth flow or material in the programme, that is, the arrangement of the frames was logical.

2. At some places difficult language was used.

3. The programme needed more practice frames.

4. Frame No. 18 was not well explained, cue and prompt were also needed in this frame.

The programme was revised according to the feedback received from one-on-one evaluation. Then the programme was tried out on a small group consisting of five IX class students. Each student was given pre-test, the programme and the post-test. Before starting the programme the group was told about the programme and the importance of the criticism. They were also asked to write any suggestions if they have. They all took 40-55 minutes to complete the programme. Finally they were interviewed by the programmer and were asked to give their suggestions. The programme was further revised according to the feedback received from small group evaluation.

Simmative Evaluation. The revised programme was given to the target population, (150 IX class students) after giving the pre-test. There were few students who asked questions when going through the programme, but when they were asked to read the matter again they seemed to understand it. All of them were able to complete the programme in time (about 2 periods of 30 minutes duration). Post-test was given the next day.

The pre-test and post-test gain shows that 72 per cent of the students achieved 80 per cent or more objectives. The majority of the students used in the summative evaluation programme showed enthusiasm in working with the programme. Many of them said that they liked the new system of instruction and wanted to have more programmes like this. The vast pre-test to post-test gains exhibited by most of the students generated a great deal of excitement and confidence. This fact alone made the whole project a worthwhile experience for the students.

Constructed Response Frame. As the word Constructed implies, no choice are presented to the student in a Constructed Response Frame. He does not select one response from many, as in the Discrimination Frame. Instead, the student constructs his own response each time. That is, he supplies the answer from his own knowledge.

The response the student constructs can take many frames. He may be asked to write or supply a word or statement, draw a diagram, or perform any other type of over action requiring a response from within his own repertory.

Examples. Frames No. 1, 2, 8,11,12,13,14,16,17,18,19,20, 21,22 and 23. The Constructed Response frame is basically a two-part structure, the set frame and at least one *practice frame*. It may be desirable to have several practice frames with each set frame.

Set Frame. Whenever the response asked for is found in the data portion of the frame, it is known as a set frame. The student may never have seen the desired response prior to reaching this frame, but he is able to supply this response simply by deducing it from the data within the frame itself.

Examples. Frames No. 1, 2,11,12,13,14,16,18 and 21.

Practice Frame. The set frame is followed by a practice frame. The practice frame gives the student a chance to practise what he has learned in the set frame.

Examples. Frames No. 8,19, 20, 22 and 23.

Discrimination Frame. As the name implies, this construction technique is used to teach student to make discrimination. A student who has been taught to make fine discriminations through the use of Discrimination Frame sequence will find that he can go beyond the programme material and approach the subject matter from any direction. For example, if taught a definition by Discrimination Frame Sequence, he will be able to define it when given the term; if given the definition, he will be able to name the item; if physically possible, he will be able to illustrate it.

Examples. Frames No. 3,4,5,6, 7,9 and 15.

Baboon Frame. In this frame the student is asked to make a choice from among four answers: Choice A, Choice B, both A and B, or neither A nor B.

Examples. Frame No. 10.

Basically, the Baboon Frame Sequence consists of three frames similar in purpose to those of the Constructed Response Frame Sequence. The first frame is a set frame; it contains enough information to enable the student to come up with the correct response when asked to respond. This frame is followed by a practice frame; here the student is asked to demonstrate, with a little prompting. In the final frame, the minimum amount of stimulus is presented to the student and a maximum response is called for.

Examples

1. A Trapezium is a figure with four sides, two of which are parallel. Place a check mark (^) before the correct statement below check only one Answer.

A. Figure A is a trapezium.

B. Figure Â is a trapezium.

C. Both Figure A and Figure Â are trapeziums.

D. Neither Figure A nor Figure Â is a trapezium.

B. Figure A is a trapezium.

2. Place a check mark (^) before the correct statement below. Check only one answer.

Fig. A *Fig. B*

(a) Figure A is a trapezium.

(b) Figure B is a trapezium.

(c) Figure A and Figure B are trapeziums.

(d) Neither Figure A nor Figure B is a trapezium. D. Neither Figure A nor Figure B is a trapezium. 3. Define "Trapezium" and draw two figures. A trapezium is a figure with four sides, two of which are parallel (or words to this effect). The example figures that you have drawn should match this definition.

Other Programmes

Linear Programme. In linear programmes, all of the students are normally required to take all of the frame.

Example. The self-learning unit, "Structure of an atom" is a linear programme.

Linear Programme strategy was used in writing this Programme due to the following reasons:

1. The topic was new for most of the students for whom it was developed, and they were needed to be provided with a lot of practice.

2. The topic of the programme was of such nature that the subject matter involved ascending order of complex skills to be learned/ therefore it was felt that all the students should be required to take all frames.

3. The type of learning task necessitated mostly constructed response answers.

Branching Programme. The word branching suggests any deviation from the straight line, A Branching Programme allows for greater differences in student abilities.

The Branching Frame Sequence Technique. "Systematics" is given in Appendix I-1. This technique presents the student with remedial information/ if necessary and permits him to take steps that are as large as his capabilities allow. A particularly adept student may go through a programme in a minimum number of steps, whereas his less able cohort may require twice that number to learn, the same amount of material. The Branching Programme offers the student alternate paths from which to choose, and the path he takes depends upon the response he makes in each frame.

Hence now you know how to develop a programmed instructional unit. Identify some topics in your field, on which you feel good programmes may be developed. Develop, evaluate and revise programmed instructional units on these topics, and share these self-learning units with your fellow teachers.

QUESTIONS

1. What is the need of programmed instructional units in effective science teaching?

2. What is programmed instruction?

3. What are the bases of Behavioural Psychology for learning through programmed instructional materials?

4. What are the advantages of programmed instruction?

5. What is the scope of integrating science practicals with theory in programmed instructional materials?

6. What is adjunct programming? What are its two types? Discuss how adjunct programming can help in better science teaching.

7. Write some strengths and weaknesses of teaching or learning science through programmed instructional materials.

8. Write down the steps for developing a programmed unit using system approach.

9. What do you mean by:

   (a) task analysis, and (b) entry behaviours? Illustrate your answer with examples.

10. Distinguish between:

    (a) formative, and (b) summative evaluations.

11. What are:

    (a) Constructed Response Frames, (b) Discrimination Frames, and (c) Baboon Frames. Illustrate your answer with some examples..

12. Differentiate between:

    (a) linear, and (b) branching programmes. Illustrate your answer with examples.

8

Teacher's Responsibility

According to F. Diesterweg, "A bad teacher teaches the truth; a good teacher teaches how to find it".

A good teacher is a congenial and conscientious person who leads an ordinary normal life. He is respected and intelligent person. He possesses a sense of humour and also an aptitude for teaching.

Another requirement for a good teacher is that he should have a high sense of principle and an aptitude for creative work and scientific curiosity.

In this chapter we will try to make a distinction between a good teacher and a good chemistry teacher we will also discuss the kind of training required to produce a good chemistry teacher.

The training of a good chemistry teacher, to a large extent, depends on the following factors:

(i) The careful selection of the candidates.

(ii) The educational process.

(iii) The efficacy of retraining programmes.

Candidate Selection

The search for potential teachers should begin while future candidates are at school. For such a selection very useful role can be played by university and college teachers in taking part, with school teachers and pupils, in chemistry competitions, evening get-togethers, science clubs etc. It is essential because only personal contacts and close acquaintance with potential teacher-training candidates can ensure success in the search of boys and girls who are sufficiently talented and gifted to became good teachers.

This process of selection should continue through out the academic career of the prospective candidate and should not end even at the end of university education.

For any one who opts to become a teacher the basic requirement is that he must be dedicated and sincerely interested in communicating knowledge. He must also be willing to undertake the ardous task of educating younger generation.

While looking for potential teachers we must ensure that only such boys and girls are selected, for being trained as chemistry teachers, who are sufficiently talented and motivated to become 'good' teachers the search for such 'good' teachers should be carried out by teachers at all levels, among secondary pupils, undergraduates and graduates. The number of teachers depend directly on the number of young people choosing this difficult career.

Training of Teachers

The preparation of a secondary school chemistry teacher involves three elements *i.e.*

(i) The academic study of chemistry,

(ii) Educational and professional studies and

(iii) School experience.

In most of the countries those who obtain their M.Sc. degree in chemistry or subjects in which chemistry plays a major role and who opt for teaching profession are trained for a year or so in special institutions (*e.g.* College of education) and awarded a degree in teaching (*i.e.* B.T., B.Ed. etc). It is a general belief that a thorough knowledge of chemistry is first and foremost for becoming a good chemistry teacher. It is also desirable for a chemistry teacher to become acquainted with those aspects of physics, biology and other natural sciences which chemists need and use.

Secondary school chemistry teachers are in short supply in most countries and even developed countries also face difficulties in recruiting specialised teachers. A serious shortage of teachers inevitably entails additional concern about quality. Not surprisingly, therefore, both these concerns, together with the need to respond to innovation in school curricula, have been important in promoting a reconsideration of the structure and content of teacher training programme in many countries.

Teachers for primary classes are usually trained in colleges of education, which may or may not be attached to the university. Teachers for senior secondary classes have followed a science course in a university.

These days there is an increasing number of university courses devoted to chemistry and education and students have to choose before going to university whether or not they wish to teach. In Malaysia B.Sc. Ed. course was introduced in four universities. Such a system with slight variations can be seen in a wide range of developed and developing countries.

In some universities an inter-linked study scheme has been introduced. *e.g.* In Yugoslavia. This type of structure is also seen in U.K. At one university in U.K., a chemistry-with-education course allows students to spend about 65% of their time working alongside chemistry undergraduates, taking the same classes and examinations. The remaining 35% of the curriculum time is used for educational studies but students still have to take a fourth-year, post-graduate course of training for the teaching profession.

In Sri Lanka some elements of chemical education have been introduced into university chemistry courses. Chemical education is also available as an optional study for a small proportion of the chemistry undergraduates in united kingdom.

In the United States, 4-year courses of concurrent study of chemistry and other sciences and of education is the common pattern. This leads to courses of approximately 60% science, 20% education and 20% general education.

In 1980's yet another approach of teacher education has emerged. It is based on Schon's notion of reflective practitioner'. This approach is committed to analysing how 'professionals think in action' and it seems to hold much promise for teacher education in general.

Recently some initiatives have been taken in United Kingdom to increase the role of schools in the teacher training process. This is quite evident in the 'articled teacher' scheme, which requires student teachers to spend most of a 2-year training period working under supervision in a school that shares responsibility for students' professional development with a training institution.

Thus we can see that the three elements of training described earlier must be inter-related: the acquisition of knowledge in the sciences; the foundation in education; and teaching methods and practice. The relative importance attached to the three parts and degree of integration between them varies from country to country.

One aspect of moving the balance in favour of methodology is the need to arrange as much teaching practice as possible. Methodology courses include not only methods of teaching but also a study and evaluation of curricula being studied at schools. The content of the methodology part of the course must also include an appreciation of assessment techniques because these will be crucial part of their pupils work and thorough training in setting questions and marking answers is needed.

Conditions for Work

Appropriate working conditions for a good chemistry teacher should include the following:

(i) Provision of graduates with certainty of employment.

(ii) Encouragement of society by giving them the esteem they deserve.

(iii) Providing them the material conditions necessary for their work, *e.g.* chemistry laboratory, library etc.

(iv) Providing them opportunities for strengthening the education and training received by them in their pre-service training.

Various ways in which school teachers can receive further training are:

By Self Improvement : It requires reading books, pamphlets and journals, consulting specialists etc. In this self improvement process T.V. programmes can contribute a lot. For success of self-improvement programme the teacher must have the time and money to buy books and pay for subscription of journals. However, secondary school teachers have seldom been found interested to utilise this opportunity of self improvement.

Organisation of Refresher Courses : Refresher courses are organised by universities for the improvement in the quality of their teachers. Such refresher courses provide an opportunity to secondary school teachers to establish working links with scientific groups, obtain first hand knowledge and become immersed in main stream of modern scientific thought.

Participation in Revision and Improvement : By such a participation teachers get an opportunity to come in close contact with each other and discuss their problems and elicit their concrete suggestions for further training.

Development of Curriculum

The type and extend of educations that training institutions can offer to their students depends on various factors. A need is felt to identify the skill areas which the trainee teacher ought to develop.

A survey was conducted in 'United Kingdom and it revealed that the seven most important skills out of a list of twenty-seven in which trainee science teachers should gain competence are:

(i) Lesson planning and preparation.

(ii) Lesson presentation.

(iii) Practical work organisation.

(iv) Teacher demonstrations.

(v) Safety in the laboratory.

(vi) Discipline and class room.

(vii) Class-questioning skills.

These areas are concerned with the short-term aim of pre-service training. Thus they aim to prepare and equip the student for first few years of class room teaching. They ignore the long-term aspects of the teacher's job. They also assume that the teacher has a mastery in his subject.

For a long-term aim such prospective teachers must be acquainted with the history, philosophy, sociology and economics of educational system.

Various curriculum development projects in teacher education have been started in different parts of the world. The aims of such projects are:

(i) Indentification of those aspects of science teaching methods which must be covered in pre-service training.

(ii) Pooling up the experience and expertise of leading teacher trainer and to share them with others.

At the university of Monash in Australia, the Australian Science Teachers Project (1976) was coordinated with science teacher educators across Australia participating. ASTEP introduced forty-seven units of activities and experiences in six sections.

1. Understanding Science (7 units)
2. Undestanding pupils (6 units)
3. Models of teaching (12 units)
4. Considering the curriculum (8 units)
5. The laboratory as a teaching resource (9 units)
6. The Australian context (5 units)

The Thai Science Teaching Project (Thai-STEP) is another such projects which aims at improving the pre-service training in all higher education institutions with teacher training responsibility across Thailand.

In United kingdom, the Nuffield Foundation provided funds for the Science Teacher Education Project (STEP). STEP pooled the ideas of over fifty science tutors in training institutions and developed and tested materials.

Such projects have been found useful even beyond their countries of origin as they provide range of activities and materials that be used selectively or modified and also provide guide lines for curriculum development in teacher education.

STEP has devised many an activities in different areas such as: aims and objectives; the nature of science and scientific enquiry; the pupil's thinking; language in science lessons; teacher-pupil-interaction; methods and techniques; resources for learning; adapting to the pupil; feedback to teacher and pupil; curriculum design; safety; laboratory design and management and the social context of science teaching.

We find that emphasis is laid on devising such activities which not only cover the identified skill areas but also give due consideration to what is likely to motivate the student teacher.

On the Job Training

It is now universally accepted that in-service education is a career-long necessity, although the means of carrying it out arc not readily available. The in-service training is quite expensive and be provided most economically.

In many countries, in-service training is a semi-voluntary activity, often taking place during school holidays. Some times such training is compulsory. In Malaysia such a training was made compulsory when the new integrated science curriculum was introduced. Similar was the situation in Thailand when IPST chemistry was introduced.

In the USSR, all teachers are required to attend refresher courses every five years.

In Yugoslavia, in-service-training, of at least 3 days annually is compulsory since 1972.

In United States, chemistry teachers are expected to earn a Master's degree or its equivalent with in their first 5 years of teaching.

In India, NCERT (New Delhi) has conducted courses for over 500 teachers to help them with new senior secondary school curriculum.

In Japan there is a provision which allows groups of teachers to study abroad for upto a month.

Similar arrangements can be found in many other countries.

Science teacher's associations are also actively participating in such in-service-training programmes. National chemical societies also make some distinctive contribution to make to the professional development of chemistry teachers. Institutions of higher education and universities are also participating in such programmes.

The following advantages accrue to the teacher by in-service-training:

(i) He can reorient himself with the latest knowlege and developments in chemistry.

(ii) He gets acquainted and acquires the latest strategies, techniques and methodology of teaching chemistry.

(iii) He can develop proper scientific attitude, temper and interests and learn scientific method for solving the problems and discovering scientific facts.

(iv) He can acquire necessary competency in motivating the students for learning chemistry and applying it to their day to day life.

(v) He can acquire necessary skills to guide his students in the form of educational, personal and vocational guidance.

(vi) He can be in a position to take active part in reconstruction and revision of curriculum, in preparation and revision of text-books, instructional material, teaching aids, evaluation scheme etc.

Interesting Method

Chemistry is a compulsory subject in curriculum of secondary schools in many a countries. It is a must for further education required by many a socially attractive occupations (medicines, engineering etc.). In view of this we should expect no problem in motivation for chemistry learning but it has been found by majority of chemistry teachers that their students consider chemistry as hard, dull and boring. To change this attitude teacher and curriculum developers made an attempt by concentrating on the materials to be learnt. Changes in curriculum occur slowly and to avaoid any frustration due to these slow changes teachers should find other ways to tackle the problem.

To make chemistry learning more interesting there should be a clear linkage between the affective and cognitive aspects of learning on the concerned culture.

Johnstone proposed the model for the situation of a learner confronted with the heavily conceptual content of chemistry. If information content does not over-load the concept understanding, perceived difficulty will be low and feeling will be positive.

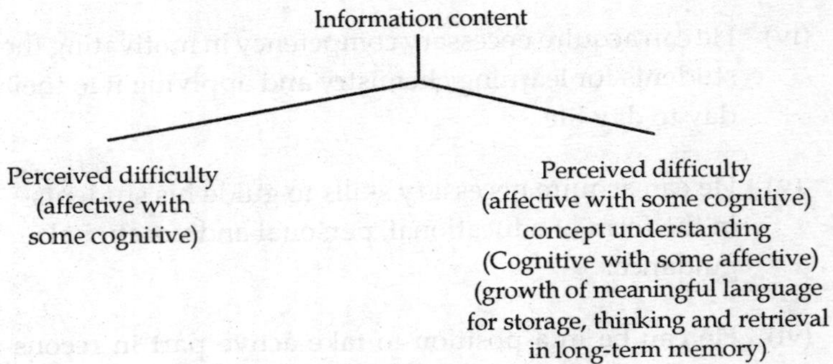

Fig. A model for learning situation confronting pupil in chemistry

For its success the teacher should explicitly explain 'Chunking' strategies. Teacher should use a consistent language and should avoid providing any unessential information. The effective use of chemistry laboratory and chemistry practicals be made by the teacher to make chemistry learning more interesting.

Significance of Laboratory

There are various types of activities that could be taken up in the chemistry laboratory. However in some countries we lack laboratory fecilities and in some others where such fecilities are available they have not been put to proper use.

Researches have proved beyond any doubts that the pupil's time in laboratories does contribute positively to their enjoyment of the subject, thus any increase in the component of a course should make it more interesting.

Karplus *et al.* developed a series of laboratory exercises for teacher in-service education that are based on Piagetian research and theory.

Gagne and White have developed a model of ways in which memory can aid or inhibit learning. Two of these postulates are more relevant for making effective use of laboratories. The first are called *images.* They are figural representation in memory of diagrames, pictures or scenes. This type of memory can be built up by chemistry teacher in the class room or laboratory.

The second are called *episodes.* There are representation in memory of part events in which the individual was personally involved.

Both *images* and *episodes* are useful aids for recall of knowledge associated with them. Generally we have those episodes which have less emotive associations but which provide a stock of concrete experiences from which meaning can be attached to new information.

Teacher should use opportunities to link the laboratory experience of the students to the learning process. He should choose images and episodes carefully and associate them with key topics in the course of study. By such an association teacher can give meaning to the abstractions of chemical knowledge.

Role and Behaviour

The duties and responsibilities of a chemistry teacher can be summarised as under:

1. He should be fully acquainted with and should have a full knowledge of school time table, the ideals of school and the social environment of the school.

2. He should be very consentious in performing his duties of teaching chemistry to various classes assigned to him.

3. He should take special interest in arranging and performing demonstration relevant to chemistry teaching in his classes.

4. He should help the students of his class to carry out practical work in the laboratory.

5. He is responsible tor organisation of chemistry laboratory, chemistry library etc.

6. He is also expected to organise various Co-curricular activities such as science fair, science exhibition, hobbies etc.

7. He is expected to help and organise the evaluation of students' progress and their achievements specifically in terms of realisation of aims and objectives of chemistry education.

8. He is also required to help in preparation and production of quality books in chemistry.

9. He is expected to select and recommend good text-books to his students.

10. He should provide active assistance in unproving chemistry curriculum.

11. He should assign appropriate and relevant home-work and assignments to his students and to check such assignments regularly.

12. He should keep a proper record of the progress of his students. Such record would be quite useful for better results.

13. He is expected to make proper use of various audio-visual aids in teaching of chemistry.

14. He is expected to help in setting up of audio-visual room in the school.

15. He is expected to help in preparation and collection of audio-visual materials and improvised apparatus.

16. He must strive hard for his own personal growth and keep himself acquainted with (i) the latest knowledge and development in the subject and methodology of teaching chemistry (ii) chemistry journals and instructional material (iii) new trends and experiments in teaching chemistry (iv) attending work-shops, summer institutes etc. (v) joining chemistry teachers association (vi) keeping himself in touch with schemes and provisions for progress of students like science scholarship, NTSE etc.

17. He should maintain a diary and make proper records in it.

18. He is expected to help in school administration and in carrying out the inspection of school specifically concerned with chemistry department.

9

Teaching Methods

By teaching chemistry we aim at bringing about a desirable behavioural changes among pupils. Teaching is thus a most difficult task and every body is not fit to be a teacher. Some persons may have a 'flair' for teaching and such persons have the ability to awaken interest and arrest the attention of the students. Some others who are not so fortunate can improve their teaching through practice if they are fully acquainted with various methods of teaching. In order to make children learn effectively, the teacher has to adopt the right method of teaching. For choosing right method for a given situation the teacher must be familiar with different methods of teaching. In this chapter an effort will be made to discuss common methods used for teaching of science.

Method of Lecturing

Lecture method is the most commonly used method of teaching chemistry. This method is most commonly followed in colleges and in schools in big classes. This method is not quite suitable to realise the real aim of teaching chemistry. In lecture method only the teacher talks and students are passive listeners. Since the students do not actively participate in this method of teaching so this method is a teacher controlled and information centred and in this method teacher works as a sole resource in class room instructions. Due to lack of participation students get bored and some of them some times may go to sleep. In this method students is provided with readymade knowledge by the teacher and due to this spoon feeding the students loses interest and his powers of reasoning and observation get no stimulus.

In this method the teacher goes ahead with the subject matter at his own speed. The teacher may make use of black board at times and may also dictate notes. This teacher oriented method in its extreme from does not expect any question or response from the students.

Advantages : It has the following advantages:

(i) It is quite economical method. It is possible to handle a large number of students at a time and no laboratory, equipment, aids, materials are required.

(ii) Using this method the knowledge can be imparted to the students quickly and the prescribed syllabus can be covered in a short time.

(iii) It is quite attractive and easy to follow. Using this method teacher feels secure and satisfied.

(iv) It simplifies the task of the teacher as he dominates the lesson for 70-85% of the lesson time and students just listen to him.

(v) Using this method it is quite easy to impart factual information and historical anecdotes.

(vi) By following this method teacher can develop his own style of teaching and exposition.

(vii) In this method teacher can easily maintain the logical sequence of the subject by planning his lectures in advance. It minimises the chances of any gaps or over-lappings.

(viii) Some good lectures delivered by the teacher may motivate, investigate, inspire a student for some creative thinking.

Disadvantages : The disadvantages of lecture method can be as under;

(i) In this method the students participation is negligible and students become passive recipients of information.

(ii) In this method we are never sure if the students are concentrating and understanding the subject matter being taught to them by the teacher.

(iii) In this method knowledge is imparted so rapidly that weak students develop a hatred for learning.

(iv) It does not allow all the faculties of the student to develop.

(v) In this method there is no place of 'learning by doing' and thus teaching by this method strikes at the very root of chemistry.

(vi) It does not take into account the previous knowledge of the student.

(vii) It does not provide for corrective feed back and remedial help to slow learners.

(viii) It does not cater to the individual needs and differences of students.

(ix) It does not help to inculcate scientific attitudes and training in scientific method among the pupils.

(x) It is an undemocratic and authoritarian method in which students depend only as the authority of the teacher. They cannot challenge or question the verdict of the teacher. This checks the development of power of critical thinking and proper reasoning in the student.

Summary : After considering various merits and demerits of method it may be concluded that this method may be suitable for teaching in higher classes (XI, XII) where we aim to cover the prescribed syllabus quickly. In these classes this method can be used successfully for imparting factual knowledge, introducing some new and difficult topics, make generalisation from the facts already known to the students, revision of lessons already learnt etc.

Teaching by this method these students of classes XI and XII will also help those students who intend to join college so that they can prepare themselves for college where lecture method of teaching is a dominant method of imparting instruction.

This method of teaching can be made more beneficial if the teacher encourages his students to take notes during the lesson. After the lesson teacher can give his students some time for asking questions and answer their queries without any hesitation. While delivering his lesson the teacher may see that the lesson is delivered in good tone, loudly and clearly. He should use only simple and understandable words for delivering his lesson. If a teacher can introduce some humour in his lesson it would keep students interested in his lesson.

Method of Demonstration

This method of teaching is sometimes also referred to as *Lecture-cum Demonstration Method*. This is considered to be a superior method of teaching in comparison to lecture method. In lecture method the teacher speaks and students listen so it is a one way traffic of flow of ideas and students are only passive listeners. This one sidedness is the major drawback of lecture method. A teaching method is considered better if both teacher and taught are active participants in the process of teaching. This particular aspect is taken care of in demonstration method.

This lecture-demonstration method is used by good chemistry teachers for imparting chemistry education in class room. By using this method it is possible to easily impart concrete experiences to students during the course of a lesson when the teacher wants to explain some abstract points. This method combines the instructional strategy of 'information imparting' and 'showing how'. This method combines the advantages of both the lecture method and the demonstration method.

In this method of teaching the teacher performs experiment before the class and simultaneously explains what he is doing. He also asks relevant questions from the class and students are compelled to observe carefully because they have to describe each and every step of the experiment accurately and draw inferences. After thorough questioning and cross-questioning the inferences drawn by the students are discussed in the class. In this way the students remain active participants in the process of teaching. The teacher also relates the outcomes of his experiment to the content of the on-going lesson. Thus while in lecture method teacher merely talks in demonstration method he really teaches.

Principle : This method is based on the principle: *Truth is that which works.*

Requirements for a Good Demonstration : For success of any demonstration following points be always kept in mind:

(i) It should be planned and rehearsed by the teacher before hand.

(ii) The apparatus used for demonstration should be big enough to be seen by the whole class. It would be much better if a large mirror is placed at a suitable angle above the teacher's table which will enable the pupils to have a view of everything that the teacher is doing while performing the experiment.

Alternately, if the class is well disciplined the teacher may allow the students to sit on the stools placed on the benches to enable them to have a better view.

(iii) Adequate lighting arrangements be made on demonstration table and a proper back ground be provided.

(iv) All the pieces of apparatus be placed in order before starting the demonstration. The apparatus likely to be used should be placed on the left hand side of the table and it should be arranged in the same order in which it is likely to be used. After an apparatus is used it should be transferred to right hand side. Only things relevant to the lesson be placed on demonstration table.

(v) Before actually starting the demonstration, a clear statement about the purpose of demonstration be made to the students.

(vi) The teacher must make sure that the demonstration-cumlecture method leads to active participation of the students in the process of learning. This he can achieve by putting well structured questions.

(vii) The demonstration should be quick and slick and should not appear to linger on unnecessarily.

(viii) The demonstration should be interesting so that it captures the attention of the students.

(ix) The teacher must be sure of success of the experiment to be demonstrated and for this he should rehearse the experiment under the conditions prevailing in the class room. However even after all the necessary precaution the experiment fails in the class room due to one reason or the other, the teacher should not get nervous instead he should make an effort to find the reasons for the failure of the experiment. Sometimes in this process a good teacher may draw very useful conclusions.

(x) No complaints about inadequate and faulty apparatus he made by the teacher. In such a situation a good teacher finds an opportunity to show his skill.

(xi) It would be much better if the teacher demonstrates those experiments which are connected with common things which are seen and handled by students in their every day life.

(xii) There should be a correlation between the demonstrations and the sequence of experiments performed by the students in their practical classes.

(xiii) For active participation of students, the teacher may call individual student, in turn, to help him in demonstration work.

(xiv) During lecture-cum-demonstration session, teacher must act like a 'showman' and a 'performer'. He should know different ways of arresting the attention of the students.

(xv) He should write, a summary of the principles arrived at because of demonstration, on the black board. The black board can also be used for drawing necessary diagrams.

Conducting Demonstration Lesson : We commonly find chemistry teachers making use of demonstration method for teaching of chemistry. The conduct of a demonstration lesson is very difficult and here we will try to discuss some of the essential steps that should be followed in a demonstration lesson.

Planning and Preparation : A great care be taken by the teacher while planning and preparing his demonstration lesson. He should keep the following points in mind while preparing his lesson:

(a) subject matter,

(b) questions to be asked;

(c) apparatus required for the experiment.

To achieve the above stated objective the teacher should thoroughly go through the pages of the text book, relevant to the lesson. After this he should prepare his lesson plan in which he should essentially include the principles to be explained, a list of experiments to be demonstrated and the type of questions to be asked from the students. These questions be arranged in a systematic order that has to be followed in the class. Before actually demonstrating the experiment be rehearsed under the conditions prevailing in the class room. Inspite of this, some thing may go wrong at the actual lesson, so reserve apparatus is often useful. The apparatus should be arranged in a systematic order on the demonstration table. Thus for the success of demonstration method a teacher has to prepare himself as thoroughly as a bride prepares herself for the marriage.

Introduction of the Lesson : As in every other subject so also in case of chemistry the lesson should start with proper motivation of the students. It is always considered more useful to introduce the lesson in a problematic way which would make students realise the importance of the topic. The usual ways in which a teachers could easily introduce his lesson is by telling some personal experience or incident, a simple and interesting experiment, a familiar anecdote or by telling a story.

A good experiment when carefully demonstrated is likely to leave an everlasting impression on the young mind of the pupil and it would set his pupils talking in school and out of it, about the interesting experiment that had been demonstrated to them in the chemistry class. This should be kept in mind not only to start the lesson but be used, on every suitable occasion, during the lesson.

It is not possible to give an exhaustive list of such interesting experiments but as an illustration we can consider the opening of soda water bottle in the class room, by the teacher, following by a direct question to his pupil, have they seen any gas coming out of the bottle? At this stage the teacher can introduce the topic of carbon dioxide.

Presentation : The method of presenting the subject matter is very important. A good teacher should present his lesson in an interesting manner and not in a boring way. To make the lesson interesting the teacher may not be very rigid to remain within the prescribed course rather he should make the lesson as much broad based as is possible. For widening of his lesson the teacher may think of various useful applications of the principle taught by him. He is also at liberty to take examples and illustrations from other allied branches of science to make his lesson interesting. The life history and some interesting facts from the life of the great chemist whose name is associated with the topic under discussion can also be cited to make the lesson interesting. Thus every effort be made to present the matter in a lively and interesting manner and a lesson should never be presented as 'dry bones' of an academic course.

Constant questions and answers should from part of every demonstration lesson. Questions and cross questions are essential for properly illuminating the principle being discussed. Questions be arranged in such a way that their answers form a complete teaching unit. Though an effort be made to encourage the students to answer a large number of questions but if students fail to answer some questions teacher should provide the answers to such questions. It is unwise to expect all the answers from the pupil and a teacher should feel satisfied if he has been able to create a desire in a student to know what he does not know.

The lesson be presented in a clear voice and the teacher should speak slowly and with correct pronunciation. He should avoid the use of any bombastic and ambiguous terms. The continuous talk is likely to monotony and to avoid it experiments be well spaced throughout the lesson.

Performance of Experiments : A good observer has been described as a person who has learned to use his senses of touch, sight, smell and hearing in an intelligent and alert manner. We want children to observe what happens in experiments and to have ample opportunities to state their observations carefully. We also want them to try to explain what happens in reference to their problem, but we want to make certain.

There is separation between observations and generalization and conclusions. We will be violating the true spirit of chemistry if we allow children to generalise from one experiment or observation.

The following steps are generally accepted as valuable in developing and concluding chemistry experiments with the children:

1. Write the problems to be solved in simple words so that every one understands.

2. Make a list of activities that will be used to solve problems.

3. Gather material for conducting experiments.

4. Work out a format of the steps in the order of procedure so that every one knows what is to be done.

5. The teacher should always try the experiment himself to become acquainted with the equipment and procedure.

6. Record the findings in ways commensurate with the maturity level and purposes of the student.

7. Assist students in making generalisations from conclusions only after sufficient evidence and experiences.

The demonstration experiment be presented by the teacher in a model way. He should work in a tidy, clean and orderly manner while demonstrating an experiment. Some of the important points to be kept in mind while demonstrating an experiment are as under:

(i) Experiments should be simple and speedy.

(ii) The experiments must work and their results should be clear and their results should be clear and striking.

(iii) Experiments be properly spaced throughout the lesson.

(iv) Keep some reserve apparatus on the demonstration table.

(v) Keep the demonstration apparatus intact till it has to be used again.

Black Board Summary: A summary of important results and principles be written on the black board. Use of black board should also be frequently made for drawing necessary sketches and diagrams. The black board summary should be written in neat, clean and legible way. Since black board summary is an index to a teacher's ability he should keep the following points in mind while writing on black board.

(i) Proper space be left between different letters and words.

(ii) Always start writing from left hand corner of the black board.

(iii) Start a new live only when the first one has extended across the black board.

(iv) Take care not to divide the words at the end of a line.

(v) Make all efforts to keep all the paragraphs and similar signs in calculations under one another.

(vi) While drawing sketches and diagrams preferably use 'single lined' diagrams.

(vii) All the diagrams drawn on the board be properly labelled.

Supervision : Students be asked to take the complete notes of the black board summary including the sketches and diagrams drawn. Such record will be quite helpful to the student for learning his lesson. Such a summary will prove beneficial only if it has been copied correctly from the black board and to make sure that students are copying the black board summary properly the teacher should check it by frequently going to the seats of the students.

Common Errors : A summary of common errors committed while delivering a demonstration lesson are given below:

(i) The apparatus may not be ready for use.

(ii) There may not be an apparent relation between the demonstration experiment and the topic under discussion.

(iii) Black board summary is not upto the mark.

(iv) Teacher may be in a hurry to arrive at generalisation without allowing sufficient time to arrive at these generalization from facts.

(v) Teacher may some times fail to ask right type of questions.

(vi) Teacher some times may use a difficult language.

(vii) Teacher some times takes to talking more which may mar the enthusiasm of the students.

(viii) Teacher may not have allowed sufficient time for recording data etc.

(ix) Teacher has not given proper attention to supervision.

Merits : Following are the merits of this method:

(i) It is an economical method as compared to purely student centred approaches.

(ii) It is a psychological method and students take active interest in teaching learning process.

(iii) It leads students from concrete to abstract situations and thus is more psychological.

(iv) It is a suitable method if the apparatus to be handled is costly and sensitive. Such an apparatus is likely to be damaged if handled by students.

(v) This method can be more safe if the experiments to be demonstrated are dangerous.

(vi) In comparison to Heuristic method, project etc., it is time saving but lecture method is too speedy.

(vii) It can be used successfully for all types of students.

(viii) In this method such experiments which are difficult for students can be included.

(ix) This method can be used to impart manual and manipulative skills to students.

Disadvantages : Some of the disadvantages of this method are as under:

(i) It provides no scope for 'learning by doing' for students as students just observe what the teacher is performing. Thus students fail to relish the joys of direct personal experience.

(ii) Since the teacher performs the experiment in his own pace, many students cannot comprehend the concept being clarified.

(iii) Since the method is not child centred so it makes no provision for individual differences. All types of students including slow learners and genius have to proceed with the same speed.

(iv) It fails to develop laboratory skills in the students. It cannot work as a substitute for laboratory work by students in which they are required to handle the apparatus themselves.

(v) It fails to impart training in scientific attitude.

(vi) In this method students many a times fail to observe many finer details of the apparatus used because they observe it from a distance.

Summary : It is thoroughly accepted that success is greater with experiments in elementary schools if they start with a real purpose, are simply done with uncomplicated apparatus, are done by children under careful direction of the teacher, and help the children think and draw valid, tentative conclusion.

This is considered as one of the best methods of teaching chemistry to secondary classes. An effort be made to involve a larger number of students by calling them in batches to the demonstration table.

Chemistry teachers should encourage more direct experimentation by children in order to help children broaden their range of fact finding skills beyond three T's-teacher, textbook, television.

Heuristic Method

Heuristic method is a pure discovery method of learning chemistry independent of teacher. The writings and teachings of H.E. Armstrong, Professor of Chemistry at the City and Guilds Institute, London, have had much influence in promoting chemistry teaching in schools. He was a strong advocate of a special type of laboratory training—heuristic training ('heuristic' is derived from the Greek word meaning 'to discover'). In Heuristic method, the student be put in the place of an independent discoverer. Thus no help or guidance is provided by the teacher in this method. In this method the teacher sets a problem for the students and then stands aside while they discover the answer.

In words of Professor Armstrong, "Heuristic methods of teaching are methods which involve our placing students as far as possible in the attitude of the discoverer—methods which involve their finding out instead of being merely told about things".

The method requires the student to solve a number of problems experimentally. To almost every one - especially children- experiments and chemistry are synonymous. Once an idea occurs to a chemist he immediately thinks in terms of ways of trying out his ideas to see if he is correct. Trying to confirm or disprove something, or simply to test an idea, is the backbone of the experiment. Experiments start with questions in order to find answers, solve problems, clarify ideas or just to see what happens. Experimenting should be part of the elementary school chemistry programme as an aid to helping children find solutions to chemistry problems as well as for helping them to develop appreciation for one of the basic tools of chemistry.

Procedure of the Method : The method requires the students to solve a number of problems experimentally. Each student is required to discover everything for himself and is to be told nothing. The students are led to discover facts with the help of experiments, apparatus and books. In this method the child behaves like a research scholar.

In the stage managed heuristic method, a problem sheet with minimum instructions is given to the student and he is required to perform the experiments concerning the problem in hand. He must follow the instructions, and enter in his note-book an account of what he has done and results arrived at. He must also put down his conclusion as to the bearing which the result has on the problem in hand. In this way he is led to reason from observation.

Essentially therefore, the heuristic method is intended to provide a training in method. Knowledge is a secondary consideration altogether. The method is formative rather than informational.

The procedures and skills in chemistry problem solving can only be developed in class rooms where searching is encouraged, creative thinking is respected, and where it is safe to investigate, try out ideas, and even make mistakes.

Teachers Attitudes : One of the most important aspects of the problem solving approach to children's development is scientific thinking in the teachers attitude. His approach should be teaching science with a question mark instead of with an exclamation point. The acceptance of and the quest for unique solutions for the problem that the class is investigating should be a guiding principle in the teacher's approach to his programme of chemistry. Teachers must develop sensitiveness to children and to their behaviour. Teachers should be ready to accept any suggestion for the solution of problem regardless of how irrelevant it may seem to him, for this is really the true spirit of scientific problem solving. By testing various ideas it can be shown to the child that perhaps his suggestion was not in accord with the information available. It can then be shown that this failure gets us much closer to the correct solution by eliminating one possibility from many offered by the problem.

In this method teacher should avoid the temptation to tell the right answer to save time. The teacher should be convinced that road to scientific thinking takes time. Children should never be exposed to ridicule for their suggestions of possible answers otherwise they will show a strong tendency to stop suggestions.

For success of this method a teacher should act like a guide and should provide only that much guidance as is rightly needed by the student. He should be sympathetic and courteous and should be capable enough to plan and devise problems for investigation by pupils. He should be capable of good supervision and be able to train the pupils in a way that he himself becomes dispensable.

Merits : This method of teaching chemistry has the following merits:

(i) It develops the habit of enquiry and investigation among students.

(ii) It develops habit of self learning and self direction.

(iii) It develops scientific attitudes among students by making them truthful and honest for they learn how to arrive at decisions by actual experimentations.

(iv) It is psychologically sound system of learning as it is based on the maxim, "learning by doing".

(v) It develops in the student a habit of diligency.

(vi) In this method most of the work is done in school and so the teacher has no worry to assign or check home task.

(vii) It provides scope for individual attention to be paid by the teacher and for closer contacts. These contacts help in establishing cordial relations between the teacher and the taught.

Limitations : Main limitations of this method are as under:

(i) It is a long and time consuming method and so it becomes difficult to cover the prescribed syllabus in time.

(ii) It pre-supposes a very small class and a gifted teacher and the method is too technical and scientific to be handled by an average teacher. The method expects of the teacher a great efficiency and hard work, experience and training.

(iii) There is a tendency on the part of the teacher to emphasize those branches and parts of the subject which lend themselves to heuristic treatment and to ignore important branches of the subject which do not involve measurement and quantitative work and are therefore not so suitable.

(iv) It is not suitable for beginners. In the early stages, the students needs enough guidance which if not given, may greatly disappoint them and it is possible that the child may develop a distaste for studies.

(v) In this method too much stress is placed on practical work which may lead a student to form a wrong idea of the nature of chemistry as a whole. They grow up in the belief that chemistry is some thing to be done in the laboratory, forgetting that laboratories were made for chemistry and not chemistry for laboratories.

(vi) The gradation of problems is a difficult task which requires sufficient skill and training. The succession of exercises is rarely planned to fit into a general scheme for building up the subject completely.

(vii) Some times experiments are performed merely for sake of doing them.

(viii) Learning by this method, pupils leave school with little or no scientific appreciation of their physical environment. The romance of modern scientific discovery and invention remains out of picture for them and the humanizing influence of the subject has been kept away from them.

(ix) Evaluation of learning through heuristic method can be quite tedious.

(x) Presently enough teachers are not available for implementing learning by heuristic method.

Summary : This method cannot be successfully applied in primary classes but this method can be given a trial in secondary classes particularly in higher secondary classes. However, in the absence of gifted teachers, well equipped laboratories and libraries and other limitations this method has not been given a trial in our schools. Even if these limitations are removed this method may not prove much useful under the existing circumstances and prevailing rules and regulations. Though not recommending the use of heuristic method for teaching of chemistry it may be suggested that at least a heuristic approach prevails for teaching of chemistry in our schools. By heuristic approach we mean that students be not spoon fed or be given a dictation rather they be given opportunities to investigate, to think and work independently alongwith traditional way of teaching.

Method of Assignment

The heuristic method is based exclusively on laboratory work where as the lecture method and demonstration method do not give any opportunity for laboratory work. For teaching of chemistry, assignment method is best suited because it involves a harmonious combination of training at the demonstration table and individual laboratory work. In this method of teaching chemistry, the given syllabus is split into well planned assignments with a set of instructions about solving the assignments. It is also possible to plan assignments based on the individual needs of the students.

Procedure : The whole of the prescribed course is divided into so many connected weekly portion or assignments. One topic is taken and a set of instructions regarding the study is drawn up. The printed page containing instructions or the assignment is handed to the pupil a week in advance of their practical work. They are then required to read the pages of the text book referred to in the assignment and write answers to a few (generally not more than three or four) questions in a note-book. The students then hand over these answers to the teacher a day before the practicals. The teacher corrects the answers. If there are a lot of mistakes in the assignments then the teacher sets the remedial and corrective assignments.

The second part of every assignment consists of laboratory work. Full instructions about laboratory work *i.e.* fitting up of apparatus, recording of results, precautions to be taken etc. On the day of the practical work the students are returned their note-books and those students whose preparatory work is found satisfactory by the teacher are allowed to proceed with the practical work.

Teaching by this method demands a lot of careful planning by the teacher and generally two out of six periods allotted to chemistry in time-table are reserved for demonstration work and remaining four for practical work. During periods reserved for demonstration work teacher gives a demonstration on a topic that is considered to be a difficult one by the pupils. These period can also be utilized by the teacher to clarify some facts which are not very clear to the pupils. For the success of assignment method the teacher should prepare a list of experiments to be demonstrated by him and another list of experiments which are to be done by the students. The success of this method mainly depends on properly drawn assignments. If the teacher keeps a progress chart he can easily distinguish between a good and an average or dull student. He can then prepare special assignments according to the needs of the student. An assignment chart may be of the following type:

The Aims : Aims of assignment method are as follows:

(i) To provide a synthesis of various methods of learning.

(ii) To provide students a training in information processing.

(iii) To develop a habit of self study among the students.

(iv) To develop scientific attitude and a habit of critical thinking among students.

(v) To expose students to various resources of leaning.

To achieve these aims the following points be kept in mind while drawing up an assignment.

(i) The assignment must be based on one textbook.

(ii) The assignment should clearly state what portion of textbook are to be read.

(iii) It should draw attention to particular points and give explanation of difficult points.

(iv) It should also indicate those portions of matter which can be omitted by the students.

(v) Questions are an essential part of the assignment and the questions be so designed that

  (a) they test whether the student has read and understood the portion assigned.

  (b) their answers are short.

  (c) their answers require diagrams to be drawn

  (d) they ask for a list of apparatus for coming laboratory work.

(vi) In each assignment the teacher should indicate portion of book dealing with the same or allied topics.

(vii) The assignment should include detailed instructions about the experiment. This portion of instructions should include.

(a) the procedure of the experiment.

(b) the method of recording results.

(c) the precautions to be observed.

(d) a diagram illustrating the set up of the apparatus.

Features of a Good Assignment

(i) It should be related to subject matter under study.

(ii) It should be concise and balanced which can be finished by student easily and quickly.

(iii) Its purpose should be clear and its objective be made known to the students.

(iv) It should be so worded that it fosters thinking and independent learning.

(v) It should be such so as to suit to the age, aptitudes and interest of the student.

(vi) It should be able to combine various methods of teaching.

Teachers Role : The teacher has to do the following for the success of assignment method of teaching.

(i) He should split up the prescribed course in chemistry into successive and progressive assignments.

(ii) He should list down the objectives for each assignment which students must achieve.

(iii) He should prepare a progress chart for each student.

(iv) He must prepare and provide a list of reference material required for each assignment.

(v) To cover up the learning gaps he should prepare remedial assignments.

(vi) He should also prepare activity sheets for laboratory work and experiments.

Merits : This method of teaching has the following advantages

(i) It provides the students an opportunity for self study.

(ii) It synthesizes various methods of teaching of chemistry and makes the learning process very effective.

(iii) It provides an opportunity to the student to learn at his own pace and thus the progress of the brighter students is not hindered by weaker students.

(iv) In this system teacher gets the central role of contingency manager and facilitator of learning. The teacher acts as a guide and interferes least in the student's work.

(v) It places more emphasis on practical work and provides students a training in skill of information processing.

(vi) It provides a feel for the scientific methods to students.

(vii) In this process the learning process can be individualized to a great extent by having differential assignment.

(viii) It provides for corrective feed back and remediation.

(ix) The progress chart with the teacher shows the progress of each student at a glance which gives the teacher an idea of a gifted and weaker students.

(x) In this process the student learns to work himself because in laboratory he is not provided with any laboratory attendant.

(xi) Habit of extra study is developed because a number of books for extra study are recommended by the teacher. Such a study helps in widening the outlook of the pupil.

(xii) Since the burden of work lies on pupil so he learns to take responsibility.

(xiii) Since the students perform experiments at their own speed so owing to their different speeds they do not perform the same experiment at the same time. Thus a large quantity of same kind of apparatus is not required.

Disadvantages : Some of the disadvantages of assignment methods are as follows:

(i) It burdens the teacher with a lot of planning and thus increases his work load to a large extent. It requires the teacher to prepare a well thought out scheme for the year before starting the method.

(ii) No source material is available in the market for assignments and preparation of assignments for different students becomes an uphill task for the teacher.

(iii) The success of method depends on the availability of rich library and laboratory facilities. It makes the method very expensive.

(iv) Before starting with this method teacher must satisfy himself that the apparatus and chemicals required for practical work are available in the laboratory. He should also satisfy himself about the availability of text books, laboratory manual, note book etc. and see that each student possesses them.

(v) Teacher should also be vigilant to see that weak students do not get a chance to copy the answers from the note books of brighter students.

(vi) Weaker students need a lot of help and guidance at individual level and it becomes an unnecessary drain on the teacher's energies.

(vii) This method is suitable only for a small group of students.

Summary : Though the method has some limitations but can be used successfully if following points are given due consideration:

(i) The teacher should prepare a well thought out plan for the year.

(ii) He should find some good resource book and use the same after necessary changes.

(iii) He should be very particular to check copying by weaker students. As remedial measures the teacher should clearly explain difficult topics and principles to the students during demonstration class and set only a limited number of questions in his assignment.

(iv) The availability of apparatus and chemicals needed for experiment be confirmed before hand.

(v) Only those students who have text book, laboratory manual and note book whose preparatory work has been found to be satisfactory be allowed to do the practical work.

(vi) A new experiment be allowed to a student when he has completed his previous experiment and has shown it to the teacher.

(vii) Students be asked to record all their observations directly in the fair note book. They should be asked to complete their practical note book in the class itself.

(viii) Teacher can provide necessary help to needy students and for this he should move from one table to another when the students are performing the experiment.

Method of Projects

This method was given by Dewey - the American philosopher, psychologist and practical teacher. The project method is a direct outcome of his philosophy. According to Dr. Kilpatrick "A project is a unit of whole hearted purposeful activity carried on preferably, in its natural setting". According to Stevenson "A project is a problematic act carried to its completion in its natural setting". According to Ballard, "A project is a bit of real life that has been incorporated into the school".

The project method is not totally new. Project equivalents are advocated for the adolescent period by Rousseau in Emile (BK-III). A project plan is a modified form of an old method called "concentration-of-studies". The main features of "concentration of studies plan" is that some subject is taken as the core or centre and all other school subjects as they arise are studied in connection with it.

Project method is based on the following principles:

(i) Learning by doing.

(ii) Learning by living.

(iii) Children learn better through association, cooperation and activity.

What is an Educational Project? Various definition of project has already been considered. A modified definition of project is given by Tomas and Long. They define it as "a voluntary undertaking which involves constructive effort or thought and eventuates into objective results".

Considering various definitions of project we may consider it as a kind of life experience which is an outcome of a craving or desire of the pupils. This is a method of spóntaneous and incidental teaching. "Learning by living" may be a better meaning of project method, because life is full of projects and individuals carry out these projects in their every day life.

The projects may broadly be classified as:

(i) Individual projects, and

(ii) Social projects.

Individual projects are to be carried out by individuals where as social projects are carried out by a group of individuals.

Steps in a Project : For completing a project we have five stages in actual practice. These are:

(i) Providing a situation.

(ii) Choosing and proposing.

(iii) Planning of the project.

(iv) Executing the project.

(v) Judging the project.

Providing a Situation : A project should arise out of a need felt by pupils and it should never be forced on them. It should be purposeful and significant. It should look important and must be interesting. For this the teacher should always be on the look out to find situation that arise and discuss them with students to discover their interests. Situations may be provided by different methods. Some such methods may include talking to students on the topics of common interest *e.g.* how did they spend their holidays, what did they see in Delhi etc.

Choosing and Proposing : From various definition of an educational project we get the same underlying ideas (a) school tasks are to be as real and as purposeful as the tasks of wider life beyond the school walks (b) they are of such a nature that the pupil is genuinely eager to carry them out in order to achieve a desirable and clearly realised aim.

According to Kilpatrick, "the part of the pupil and the part of the teacher, in most of the school work, depends largely on who does the proposing". The teacher should refrain from proposing any project otherwise the whole purpose of the method would be defeated. Teacher should only tempt the students for a particular project by providing a situation but the proposal for the project should finally come from students. The teacher must exercise guidance in selection of the project and if the students make an unwise choice, the teacher should tactfully guide them for a better project. The essentials of a good project are:

(i) It should have evident worth for the individual or the group that undertakes them.

(ii) The project must have a bearing on a great number of subjects and the knowledge acquired through it may be applicable in a variety of ways

(iii) The project should be timely

(iv) The project should be challenging.

(v) The project should be feasible.

It is for the teacher to see that the purpose of the project is clearly defined and understood.

Planning : The students be encouraged by the teacher to plan out the details of the project. In the process of planning teacher has to act only as a guide and he should give suggestions at times but actual planning be left to the students.

Execution : Once the project has been chosen and the details of the project have been planned, the teacher should help the students in executing the project according to the plan. Since execution of a project is the longest step in the project method so it need a lot of patience on the part of the students and the teacher. During this step the teacher should carefully supervise the pupils in manipulative skills to prevent waste of materials and to guard accidents. The teacher should assign work to different students in accordance with their tastes, interests, aptitudes and capabilities. Teacher should see that every member of the group gets a chance to do some thing. Teacher should constantly check up the relation between the chalked out plans and the developing project and as far as possible 'at the spot' changes and modification be avoided. However if such changes become unavoidable these should be noted and reasons explained for future guidance.

Evaluation : The evaluation of the project should be done both by the pupils and the teacher. The pupils should estimate the qualities of what they have done before the teacher gives his evaluation. The evaluation of the project has to be done in the light of plans, difficulties in the execution and achieved results. Let the students have self criticism and look through their own failings and findings. This step is very useful because as a result of the project, the pupils can know the values of the information, interest, skills and attitudes that have been modified by the project.

Record : A complete record of the project be kept by the students. The record should include every thing about the project. It should include the proposal, plan and its discussion, duties allotted to different students and how far were they carried out by them. It should also include the details of places visited and surveyed, maps etc. drawn, guidance for future and all other possible details.

Role of Teacher

(i) In project method of teaching the role of a teacher is that of a guide, friend and philosopher.

(ii) He helps the students in solving their problems just like an elder brother.

(iii) He encourages his students to work collectively, amicably in the group.

(iv) He also helps his students to avoid mistakes.

(v) He makes it a point that each member of the group contributes some thing to the completion of the project and in this process helps the shy and Weaker students to work along with their classmates.

(vi) If the students face failure during execution of some steps of the project the teacher should not execute any portion of the project but should only explain to his students the reasons of their failure and should suggest them some better methods or techniques that may be used by them next time for the success of the project.

(vii) During the execution step teacher also learns something.

(viii) Teacher should always remain alert and active during execution, step and see that the project goes to completion successfully.

(ix) During execution of the project teacher should maintain a democratic atmosphere.

(x) Teacher must be well read and well informed so that he can help the students to the successful completion of the project.

The Merits

(i) It is a method of teaching based on psychological laws of learning. The education is related to child's life and he acquires it through meaningful activity.

(ii) It imbibes the spirit of cooperation as it is a cooperative venture. Teacher and students join in the project.

(iii) It stimulates interest in natural as also man made situations. Moreover the interest is spontaneous and not under any compulsions.

(iv) The method provides opportunities for pupils of different tastes and aptitudes within the frame work of the same scheme.

(v) It upholds the dignity of labour.

(vi) It introduces democracy in education.

(vii) It brings about a close correlation between a particular activity and various subjects.

(viii) It is a problem solving method and places very less emphasis on cramming or memorising.

(ix) It helps to inculcate social discipline through joint activities of the teacher.and the taught.

(x) A project can be used to arouse interest in a particular topic as it blends school life with outside world. It provides situations in which the students come in direct contact with their environment.

(xi) It develops self confidence and self discipline.

(xii) A project tends to illustrate the real nature of the subject.

(xiii) A project affords opportunity to develop keenness and accuracy of observation and produces a spirit of enquiry.

(xiv) It puts a challenge to the student and thus stimulates constructive and creative thinking.

(xv) It provides the students an opportunity for mutual exchange of ideas.

(xvi) This method helps the children to organise their knowledge.

The Drawbacks

(i) Projects require a lot of time and this method can be used as a part of science work only.

(ii) Though the method provides the student superficial knowledge of so many things it provides insufficient knowledge of some fundamental principles.

(iii) In the project planning and execution of the project the teacher is required to put in much more work in comparison to other methods of teaching.

(iv) The teacher has been assumed as master of all subjects which is practically not possible.

(v) Good text books on these lines have not yet been produced.

(vi) It is an expensive method is it involves tours, excursions, purchase of apparatus and equipment etc.

(vii) The method of organising instructions is unsystematised and thus the regular time table of work will be upset.

(viii) The method may fit those who cannot listen but it is very questionable if it has the same value for those who can listen.

(ix) The method leaves a gap in pupils knowledge.

(x) It under estimates man's power of imagination which enables him to savour the full experience of another without the necessity of undergoing the experience himself.

(xi) Some times the projects may be too ambitious and beyond pupils capacity to accomplish.

(xii) Larger projects in hands of an unexperienced teacher lead to boredom.

(xiii) The education given by projects is likely to emphasise relationships in breadth than in depth.

Summary : The project method provides a practical approach to learning of both theoretical and practical problems. If it is difficult to follow this method of teaching it would be better at least not to ignore the spirit of this method.

This method has been found to be more suitable for primary and middle classes and is of restricted use for high and higher secondary classes. This method may be tried alongwith formal class room teaching without disturbing the school time-table. With this in view some projects may be undertaken by the students to be completed on certain fixed days of a week. Alternately first half of the day may be devoted to class room teaching and the project work be carried out in the remaining half day. To help solve the problem of fund's shortage such projects be choose which are self-supporting or the projects selected be such that their final products can be sold to partially support the funds. Some such projects are improvising chemistry apparatus, etc. Costly projects should be avoided. As it is not suitable for drill and continuous and systematic teaching, it is not very desirable to use it freely.

Concentric Method

This is a system of organising a course rather than a method of teaching. It is therefore better to call it *concentric system* or *approach*. It implies widening of knowledge just as concentric circles go on extending and widening. It is a system of arrangement of subject matter. In this method the study of the topic is spread over a number of years. It is based on the principle that subject cannot be given an exhaustive treatment at the first stage. To begin with, a simple presentation of the subject is given and further knowledge is imparted in following years. Thus beginning from a nucleus the circles of knowledge go on widening year after year and hence the name concentric method.

The Procedure : A topic is divided into a number of portions which are then allotted to different classes. The criterion for allotment of a particular portion of the course to a particular class are the difficulty of portion and power of comprehension of students in that age group. Thus it is mainly concerned with year to year teaching but its influence can also be exercised in day-to-day teaching. Knowledge be given today should follow from knowledge given yesterday and should lead to teaching on following day.

The Merits

(i) This method of organisation of subject matter is decidely superior to that in which one topic is taken up in particular class and an effort is made to deal with all aspects of the topic in that particular class.

(ii) In provides a frame work from science course which is of real value to students.

(iii) The system is most successful when the teaching is in hands of one teacher because then he can preserve continuity in the teaching and keeps his expanding circle concentric.

(iv) It provides opportunity for revision of work already covered in a previous class and carrying out new work.

(v) It enables the teacher to cover a portion according to receptivity of learner.

(vi) Since the same topic is learnt over many years so its impressions are more lasting.

(vii) It does not allow teaching to become dull because every year a new interest can be given to the topic. Every year there are new problem to solve and new difficulties to overcome.

The Drawbacks : For the success of this approach we require really capable teacher. If a teacher becomes over ambitious and exhausts all the possible interesting illustrations in the introductory year then the subject loses its power of freshness and appeal and nothing is left to create interest in the topic in subsequent years.

In case the topic is too short or too long then also the method is not found to be useful. A too long portion makes the topic dull and a two short portion fails to leave any permanent and lasting impression on the mind of the pupil.

Summary: It is a good method for being adopted for arranging the subject matter. It should be kept in mind, by the organisers, while organising the subject matter that no portion is too long or too short. It would also be much useful if the same teacher teaches the same class year after year so that he can reserve some illustrative examples for each year and thus can maintain the interest of the students in the topic.

Method of Unit

It is one of the latest methods in the field of education. It involves pupils more actively in learning process.

Different authors define unit in a different way. Hanna, Hageman, Potter define it as, "a unit is a purposeful learning experience that is focussed on some socially significant understanding which will modify the behaviour of learner and adjust him to adjust to a life situation more effectively".

However all the definitions of unit imply that it possesses the following characteristics:

(i) It is an organisation of activities around a purpose.

(ii) It has significant content.

(iii) It involves students in learning process.

(iv) It modifies the students behaviour to such an extent that he can cope with new problem and situations more competently.

Types of Units : Mainly the units may be classified as:

(i) Subject matter units.

(ii) Experience units.

(iii) Resource units.

The teaching of chemistry can be carried out in a better way and it is better understood and appreciated by the students if it is taught as units of immediate interest to the pupils. Such units may be (i) life centred (ii) environment centred (iii) life and environment centred.

For this The Tara Devi Seminar (1956) recommended the following:

Life Centred Units

1. The air we breathe.
2. The water we use.
3. The food we eat.
4. The clothes we wear.
5. Our mineral resources.
6. The universe we live in.

Environment-centred Units

1. The atmosphere.
2. Water, a vital need of life.

3. The earth surface.

4. Civilization and the use of metals.

Environment of Life-centred Units

1. Using mineral resources for better living.

2. The weather and what we can do about it.

3. Chemistry for our homes.

Interesting lessons can be developed on 'Air', 'Water' etc. These can be used for teaching of hydrogen, nitrogen, water, carbon dioxide etc.

Essentials of a Good Unit

(i) It should deal with a sizeable topic.

(ii) It should emerge out of students past experiences and should lead to broader interests.

(iii) It should be of appropriate difficulty in terms of child's understanding, interest.

(iv) It should provide scope for using a variety of materials and activities like community resources, audiovisual materials etc.

(v) It should allow use of sufficient amount of books and other learning materials.

(vi) Units should be such as to draw materials from several fields so that children may develop richer insight into human relationships and processes.

(vii) It should be functional and should be in accordance with the maturity level of the learner.

The Merits

This method of teaching has the following advantages:

(i) It brings about a closer integration between various branches of science.

(ii) It makes subject matter more interesting and realistic.

(iii) It provides a better understanding of the environment and life.

(iv) It focuses attention on significant facts and avoids confusion.

(v) The unit because of its flexibility provides facility in adopting instructions to individual's differences.

The Limitations

(i) This method cannot be used if the teacher is required to complete some prescribed course in a specified time.

(ii) There are only a few teachers who are so widely read that they can introduce material and illustration from various branches of science while keeping before their students one central topic.

Unit method or *topic method* is a varied slightly in America. In American schools the teacher announces one topic and the students are asked to say what they already know about it. Then the topic is discussed in a question and answer session and those questions which no member of the class could answer are noted down for investigation. From this list of questions, such questions as are considered as too difficult for a particular class are eliminated by the teacher and the remaining questions are arranged in a planned manner for answers. These questions are then dealt within the class according to the plan. The great thing about such a course is that boys feel that it is their course and not some thing thrust upon them by authority.

Method of History

Some teachers prefer to develop a subject by following the stages through which the subject has passed during its course of development from its early beginnings. This type of teaching has a fascination which appeals to pupils. Various science subjects such as Chemistry, Bacteriology etc. Which have an interesting historical background can be taught successfully by such a technique. It is possible to develop a topic starting from its early history and the various stages through which it developed before attaining the modern shape.

Chemistry, has a very interesting history and the works of Priestley, Lavoisier, Davy, Black and Dalton etc. can be given this type of treatment. The gradual development of atomic theory can be unfolded gradually by this method which will be quite interesting.

Through such a treatment may not be possible for all the topics but an occasional resort to such a treatment has its own uses.

Method of Discussion

This method is found quite suitable for those topics in chemistry which cannot be easily explained by demonstration or other such techniques. The discussion may be about a certain specimen or model or chart.

In this method the topics for discussion is announced to the students well in advance. The teacher gives a brief introduction about the contents of the topic and them suggests to his students various reference books, text books and other books. Students are then required to go through the relevant pages of these books and come prepared for a discussion of the topic on a specified day. During actual discussion period teachers poses a lew problems and thus provides the necessary motivation. The students are then asked to answer the question one by one and when ever thinks fit advises some students not to go out of the scope of a particular question or topic under consideration. This check is essential otherwise immature students may go out of the scope of the topics.

Following points if kept in view will help make the discussion successful:

(i) The topics for discussion should be of common interest of students.

(ii) Teacher should establish a favourable atmosphere in the class before starting the discussion.

(iii) Teacher should see that every one participates in the discussion. The whole essence of discussion is "Thinking together".

(iv) The teacher should talk to the bare minimum and also should not allow any one student to dominate the whole discussion.

(v) It is for teacher to see that the discussion remains a discussion and it does not change into a debate.

(vi) Teacher should keep a check on answers of the students and should not allow a student to go beyond the scope of a topic under discussion.

(vii) Teacher has to maintain discipline and he should see that only one student speaks at a time.

It is a combination of two methods. To be able to understand this combination it is necessary to understand them separately.

Method of Induction

In this method one is led from concrete to abstract, particular to general and from complex to general rule. In this method we prove a universal law by showing that if it is true in a particular case it is also true in other similar cases.

This method has been found to be quite suitable for teaching of chemistry because most of the principles of chemistry or the conclusions are results of induction. This process of arriving at generalisation can be illustrated as under:

Illustration. Take a piece of blue litmus paper and dip it in a test tube containing hydrochloric acid, observe the change in colour. (It turns red.)

Take another piece of blue litmus paper and dip it in a test-tube containing nitric acid. Observe the change in colour. (It turns red.)

Repeat the experiments with other acids in different test tubes (*e.g.* oxalic acid, acetic acid etc.). (In each case blue litmus turns red.)

From the above experiments we can make a generalisation that *acids turn blue litmus red.*

The Merits

(i) It helps understanding.

(ii) It is a scientific method.

(iii) It develops scientific attitude.

(iv) It is a logical method and develops critical thinking and habit of keen observations.

(v) It is a psychological method and provides ample scope for students activities.

(vi) It is based on actual observations, thinking and experimentation.

(vii) It keeps alive the students interest because they move from known to unknown.

(viii) It curbs the tendency to learn by rote and also reduces home work.

(ix) It develops self-confidence.

(x) It develops the habit of intelligent hard work.

The Drawbacks

This method suffers from the following limitations:

(i) It is limited in range and cannot be used in solving and understanding all the topics in chemistry.

(ii) The generalization obtained from a few observations are not the complete study of the topics. To fix the topic in the mind of the learner a lot of supplementary work and practice is needed.

(iii) Inductive reasoning is not absolutely conclusive. The generalization has been done from the study of a few (three or four) cases. The process thus establishes certain degree of probability which can be increased by increasing the number of valid cases.

(iv) This method needs a lot of time and energy and thus it is time consuming and laborious method.

(v) This method is not found to be suitable in higher classes because some of the unnecessary details and explanations may make teaching dull and boring.

(vi) The use of this method should be restricted and confined to understanding the rules in the early stages.

(vii) This method may be considered complete and perfect only if the generalization arrived at by induction can be verified through deductive method.

Method of Deduction

Deductive method is opposite of inductive method. In this method the learner proceeds from general to particular, from abstract to concrete. Thus in this method facts are deduced or analysed by the application of established formula or experimentation. In this case the formula is accepted by the learner as a duly established fact.

In this method teacher announces the topics of the day and he also gives the relevant formula/rule/law/principle etc. The law/formula is also explained to the students with the help of certain examples which are solved on the black board. From these students get the idea of use or application of the concerned law/principle/formula. Then the problems are given to the students who solve the problems following the same method as explained to them earlier by the teacher. Students also memorise the results for future application.

Following example illustrates the procedure:

Principle: Cooling is caused by evaporation.

Confirmation by Application: It can be confirmed by numerous application, such as, by wearing wet clothes, observing feeling after taking bath, by applying alcohol on your hand etc.

The Merits

(i) It is short and time saving and so this method is liked by authors and teachers.

(ii) It is quite a suitable method for lower classes.

(iii) It glorifies memory because students are required to memorise a large number of laws, formulae etc.

(iv) For practice and revision of topic it is an adequate and advantageous method.

(v) It supplements inductive method and thus completes the process of inductive-deductive method.

(vi) It enhances speed and efficiency in solving problems.

The Limitations

(i) It is not a scientific method because the approach of this method is confirmatory and not explanatory.

(ii) It encourages rote memory because pure deductive work requires some law; principle formula for every type of problem and it demands blind memorisation of large number of such laws/formulae etc.

(iii) Being an unscientific method it does not impart any training in scientific method.

(iv) It causes unnecessary and heavy burden on the brain which may some times result in brain fag.

(v) In this method memory becomes more important than understanding and intelligence which is educationally not sound.

(vi) It is an unpsychological method because the facts and principles are not found by the students themselves.

(vii) In this method students cannot become active learners.

(viii) It is not suitable for development of thinking, reasoning and discovery.

Summary : A careful consideration of merits and limitations of these two methods leads us to conclude that Inductive Method is the fore-runner of Deductive Method. For effective teaching of chemistry, both inductive and deductive approaches should be used because no one is complete without the other. Induction leaves the learner at a point where he cannot stop and the after work has to be done and completed by deduction. Deduction is a process that is particularly suitable for final statement and induction is most suitable for exploration fields. Induction gives the lead and deduction follows. In chemistry if we want to teach about *composition of water* then its composition is determined by a endiometer tube (inductive process) and confirmed by the process of electrolysis of water (deductive process).

Method of Science

This method of teaching of chemistry is based upon the process of finding out the results by attacking a problem in a number of definite steps. It is possible to train the students in scientific method. In this method student is involved in finding out the answer to a given scientific problem and thus actually it is a type of discovery method.

Fitzpatrick defines science as, "science is a cumulative and endless series of empirical observations which result in the formation of concepts and theories, with both concepts and theories being subject to modification in the light of further empirical observation. Science is both a body of knowledge and the process of acquiring and refining knowledge".

Considering this definition of science it becomes imperative that the students be exposed to the scientific way of finding out. Scientific method of teaching helps to develop the power of reasoning, application of scientific knowledge, critical thinking and positive attitude, in the learner.

This method proceeds in the following steps:

(i) Problem in identified.

(ii) Some hypothesis are framed and these are proposed for testing.

(iii) Experiments are then devised to test the proposed hypothesis.

(iv) Data is collected than observations and the collected data is then interpreted.

(v) Finally conclusions are drawn to accept, reject or modify the proposed hypothesis.

Scientific method is therefore a well sequenced and structured method for finding the results through experiments.

Role of Teacher : For the success of scientific method the role of teacher is very important. He should act as a co-investigator along with students and must also find sufficient time and have patience to attend to students' problems. Under the proper guidance of the teacher the science laboratory should become the hub for implementations of this method.

The Merits : Scientific method has following advantages:

(i) Students learn chemistry of their own and teacher works only as a guide.

(ii) It helps students to become real scientists as they learn to identify and formulate scientific problems.

(iii) It provides to students a training in techniques of information processing

(iv) It develops a habit of logical thinking in the students as they are required to interpret data and observations.

(v) It helps to develop intellectual honesty in students.

(vi) It helps the students to learn to see relationships and pattern among things and variables.

(vii) It provides the students a training in the methods and skills of discovering new knowledge in chemistry.

The Limitation : Some important limitations of the method are as under:

(i) It is a long, drawn out and time consuming method.

(ii) It can never become a full fledged method of learning chemistry.

(iii) Due to lack of exposure to this method most of the chemistry teachers fail to implement it successfully.

(iv) This method is suitable only for very bright and creative students.

Method of Problems and Solutions

In this method of teaching chemistry the student is required to solve a problem by an experimental design making use of his previous knowledge.

In chemistry problem solving has been presented by Ashmore, Frazer and Casey. They define problem solving as a result of the application of knowledge and procedure to a problem situation and propose four stages:

(i) definition of the problem,

(ii) selection of the appropriate information,

(iii) putting together the separate pieces of information, and

(iv) evaluation of the solution.

Many other authors in the field have taken up different approaches and some have given even 20 to 30 steps in the problem solving approach for teaching of chemistry.

Inspite of the enthusiasm for designing stages in problem solving, there is hardly any impact an chemical education.

Difficulties in Problem Solving in Chemistry : There were a many fold difficulties, some of these are:

(i) To formulate problem which will be real but still have only one solution.

(ii) To select such problems which include relatively simple chemistry.

(iii) To get teachers interested in research, which is considered as the best method of developing problem—solving skills.

(iv) To construct problems for chemistry teaching and learning which will be of help not only at tertiary, but also at secondary level.

To construct problems for teaching problem solving skills the following procedure has been developed.

(i) Select a paper from research journal. (This paper should not be easily available to students).

(ii) Take out the research data from the selected paper and handed over to the students as a list of experimental results.

(iii) Students are then asked to attempt to solve the problem as well as to design their problem—solving network.

(iv) Students network are compared and analysed, to identify the reasons for success or failure.

(v) This forms the basis of discussion on problem solving, problem solving network and problem solving skills in which teacher and students participate.

(vi) In some cases the problem is given to different groups and results are compared for evaluation.

In well structured examples, in which students use more or less the same network independently, efforts are made to develop the similar problems for lower level. This is known as 'funnel approach' (Fig.)

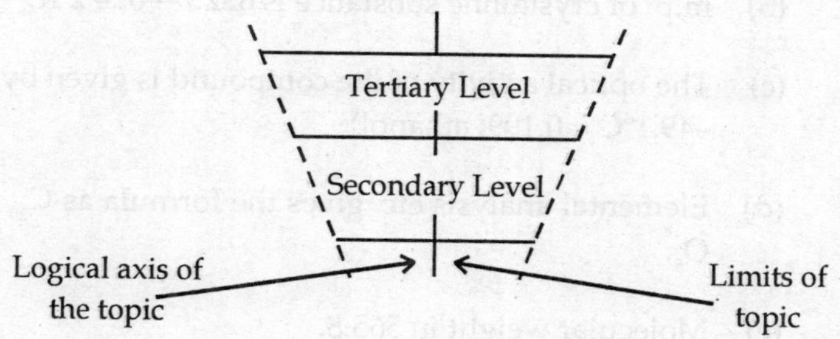

Fig. The Funnel Approach

To illustrate it the problem selected is from chemistry of natural products.

Selection of Research Paper. The paper on "The structure of viscumamide, a New Cyclic Peptide Isolated from *Viscum album* Linn. Var. *Coloratum* Ohwi" was selected.

This paper was selected because

(a) of its high quality

(b) of its relevance to undergraduate course

(c) of its demanding relatively simple knowledge of chemistry

(d) of its having a sufficient amount of experimental data.

Useful Data that was Available in the Paper was

(a) The unknown compound was isolated from the neutral fraction of the methanol extract of *Viscum album* Linn. Var. *Coloratum* Ohwi by column chromatography on activated alumina.

(b) m.p. of crystalline substance is 622.7—624.2 K.

(c) The optical activity of the compound is given by $[\alpha]_D^{25}$= –49.1°C = 0.199, ethanol)

(d) Elemental analysis etc. gives the formula as $C_{30} H_{55} N_5 O_5$.

(e) Molecular weight in 565.8.

(f) I.R. Spectra gives bands (absorption) at 3315, 3050, 1658, 1530 cm^{-1}. No absorption bands are obtained which may indicate the presence of free amino group or a carboxyl group.

(g) Potentiometric titration of 0.01 m HC1 with 0.1 m NaOH gives the same curve as that of a blank titration.

(h) It gives no reaction with ninhydrin.

(i) It is sparingly soluble in most of the common solvents.

(j) Hydrolysis in 6 m HC1 at 383.2 K for 120 hours gave two products, 60.6% of substance A and 39.4% of substance B.

(k) Both A and B have the molecular formula $C_6 H_{13} O_2 N$ and both are optically active.

$[\alpha]_D^{27}$ = +12.1° (for A) (C in 20% HC1 = 0.40) = + 34.8° (for B) (C in 20% HC1 = 0.39)

(l) Products of A and B form esters with butan-1-ol saturated with HC1.

(m) Products A and B from amides with trifluoroacetic anhydride in dichloromethane.

Problem : Try to hypothetize what the isolated substance would be.

Design of Networks : Students were free to design their network and it was observed that the networks of successful students were almost similar.

Analysis of Students' Results : Various groups were involved in solving the problem. Following generalisation were made.

(i) Since most of the groups were able to give good results, this should be included in every curriculum to develop problem-solving skills.

(ii) Some students were careless and they 'lose' pieces of information.

(iii) A large number of students had no patience to build up the network and got lost in speculation.

(iv) Most of the students forgot to check their solution by comparing it with the given experimental data.

Development of a Similar Example for Secondary Level : The following example was developed and tried out in secondary schools:

Problem : Hydrolysis of proteins gives a solid compound A which dissolves in water. In electrolysis it migrates towards the cathode or anode—depending on pH. It is not optically active. On heating it yield another solid B having a molecular mass 114. What is A?

Students Results : In most cases students could reach only second level. Coming to linear conclusions. 50% of students could reach the hypothesis that A may be glycine but most of the students failed to check if substance B could be cyclic peptide.

Conclusion

(i) Research papers are good source for construction of problems for teaching chemistry through problem-solving method.

(ii) Students be asked to solve the problem as also to design the problem-solving network.

(iii) Instead of individuals, individualised groups be encouraged for problem-solving. It gives more confidence to the students.

(iv) Original nętwork that leads to problem-solving should be encouraged.

(v) Positive and negative results should be carefully analysed and discussed with students.

(vi) A speculative approach should be discouraged.

(vii) Similar types of problems for different school levels be designed.

Learning Chemistry by Pattern Recognition : This method of chemistry learning is based on the following hypothesis. In chemistry we find the existence of periodic system. It is a recognised pattern of a number of facts about elements and their relationships.

Similar pattern of compounds exists in matter. However to reach a bigger pattern of compounds we should first aim at subsystems.

The main idea of learning chemistry *by pattern recognition* is to encourage the students to select a specific field of compounds or reactions or properties to search for their characteristics and and to try to construct patterns and check them.

This is illustrated by taking example from the course of secondary level.

Example 1. **Organic Compounds of Oxygen**

Students learn organic compounds only in fragments however they may be presented as a chain of knowledge.

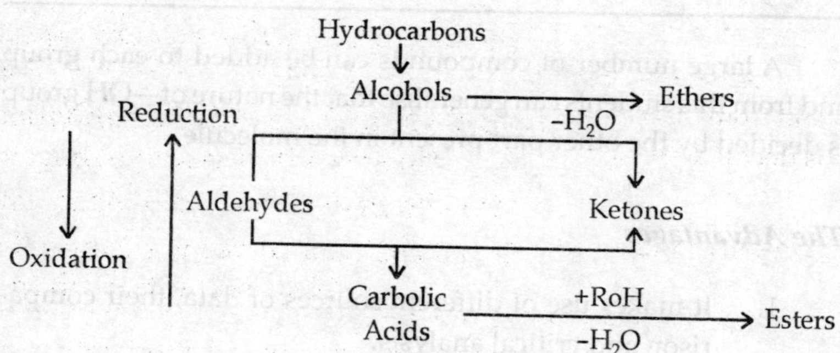

This chain of knowledge is a pattern. It is repeated at a higher level in chemistry of carbohydrates.

Considering only a fragment of carbohydrate molecule *(i.e.* dioxycarbohydrates) we can find some analogy with hydrocarbons. It helps the students in learning more about the carbohydrates and their inter-relationship with oxygen compounds.

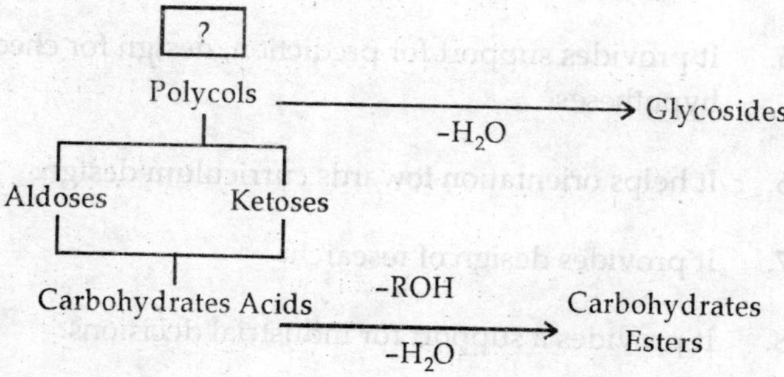

Example 2. The Chemistry of—OH group

The compounds containing -OH group can be classified as acidic, neutral or basic.

Acidic	*Neutral*	*Basic*
Phenol	Methanol	Sodium hydroxide
C_6H_5–OH	CH_3-OH	NaOH

A large number of compounds can be added to each group and from this students can generalise that the nature of—OH group is decided by the other part present in the molecule.

The Advantages

1. It makes use of different sources of data, their comparison and critical analysis.
2. It motivates students for research and data organisation.
3. It helps organise learning towards higher cognitive levels.
4. It develops the ability to organise the data, to find an classify parameters, to search for regularities, systems and patterns.
5. It provides support for prediction, design for checking hypotheses.
6. It helps orientation towards curriculum design.
7. It provides design of research.
8. It provides a support for industrial decisions.

The Drawbacks

1. It is too vague.

2. In the absence of sufficient data students might end in speculation.

3. It involves danger of over simplification and also of 'model thinking'.

Summary : Pattern recognition can combine teaching and learning situations with research design and is also helpful in taking industrial decisions. It is one of the most efficient methods in teaching of chemistry.

To avoid the drawbacks of the method the process of pattern recognition should be considered more important than the pattern produced. Critical evaluation of a pattern be taken up carefully. Every pattern design has to be faced not only with facts and their relationships, but in a dialectic approach, in which the recognition of their changing nature is also essential.

Selection of Method

In the previous pages a number of methods for teaching of science have been discussed. Some of them have been recommended for use, some have been disapproved and some have been recommended for use with caution. Out of the methods available a choice is not entirely left to the whims of the teacher but has to be made by the teacher in the light of facilities available and nature of work to be done. This does not mean that a teacher may select any one method and then cling to it lavishly throughout his service or even an entire academic session. This is a great mistake because each method has its own merits. Our preference for any one of the methods deprives us of the merits of other method. A good teacher should therefore try to imbibe the good qualities of all methods instead of depending on any one method. The teacher should keep himself on the right side of every method. The best method of a teacher is his own individualised and personalised method which is the result of his varied and long experience in teaching. Some of the points which a teacher should keep in mind are as under:

(i) Heuristic approach be used to start a lesson. Thus the lesson be introduced in a problematic way so that the students feel that they are going to learn some thing really useful and worth learning.

(ii) He should choose a pupil—dominated method in preference to a teacher dominated method.

(iii) He should have a bright manner of presentation and should illustrate his lesson with experiments, pictures, charts, diagrams etc. specimens and models are preferred for illustration.

(iv) Teaching should be made a cooperative enterprise. Teacher should give maximum opportunity of participation to the students so that they feel that their active participation in quite important for the solution of problem and successful growth of the subject.

(v) Teacher should made all possible efforts to properly correlate the topic in hand with other subjects.

(vi) Teacher should avoid the use of difficult phrases, scientific expressions and lengthy definitions.

(vii) Through Heuristic approach dominates that the historical method of teaching be utilized at places and the lives and achievements of famous scientists be told to the students. These are a source of inspiration to the students.

(viii) Instructional method and plans must be flexible. In a lesson if, in addition to planned illustrations and experiments students want some more experimental evidence then the teacher should make all possible efforts to satisfy the students.

(ix) After a constant use of some method teacher can break the monotony by using project method and laboratory method.

Thus we conclude that no single method could be the best method and a good teacher will have to evolve his own individual method consisting of good points of all the methods. He will never become a slave to any method and will remain a true master of all of them.

10

Devices for Teaching

The teaching aids are required by a chemistry teacher, like teachers of other subjects, for effective teaching of subject and to realise various objectives of teaching chemistry. Teaching aids help the teacher to communicate with his students in more desirable and effective way. Some barriers of communication can be overcome by using special aids appealing to the senses of the receiver alongwith managing the communication along certain principles. Class room instructions or teaching a curriculum transaction is also a special kind of communication and it is helpful in achieving the instructional goals of a course of study. Effective communication requires a mastery of managerial skills of handling various teaching aids like audio-visual aids, visual aids, audio aids, activity aids etc.

Various Categories

Teaching aids are classified, for convenience of study, into the following categories:

(i) Audio aids (ii) Visual aids (iii) Audio-visual aids (iv) Activity aids.

Examples of various types of teaching aids generally used to make class room teaching of chemistry more effective are given below:

Audio Aids : In this type of aids fall the teaching aids like radio, tape recorder etc. This type of aids help the process of learning as they help the learner to acquire knowledge through his auditory senses.

Visual Aids : This type of aids are very common *e.g.* charts, pictures, models, film strips etc. These aids the learner to acquire the learning experiences through his visual senses.

Audio-visual Aids : These are sensory aids which help to make teaching concrete, effective and interesting. Examples of this type of aids are television, motion picture, video films, living objects etc. By use of these aids we provide the learner an opportunity to utilise both his auditory and visual senses for gaining the desired learning experiences.

Activity Aids : In this type of aids we include all those teaching aids in which the learner is required to be engaged in some useful activity, *e.g.*

(a) Excusions and visits.

(b) Exhibitions and fairs.

(c) Experimentation in the laboratory and work-shop.

The Significance

Teaching aids make the teaching-learning process interesting and more meaningful as we are required to make use of our senses. While commenting on the desirability of making use of ones senses the Indian Education Commission has remarked, "for acquiring right and proper knowledge and experiences regarding the objects and processes must be gained through one's senses".

The importance of teaching aids can be summarised as under:

(i) Teacher can win the interest and attention of the pupils by making use of teaching aids.

(ii) They are effective motivating agents.

(iii) They help to bring clarity to the subject matter.

(iv) They same time and energy of the students and teachers and make learning more effective and durable. A fact, principle or phenomenon that cannot be understood properly with verbal explanation or experience can be easily comprehended by use of teaching aids. In this way the time and energy of both the students and teacher is saved.

(v) Proper use of teaching aids helps to develop in the pupils scientific attitudes and provide them with a training in scientific method.

(vi) They provide the pupils with the first hand experience by looking at concrete things and actual demonstrations.

(vii) They provide a solution to a number of educational and administrative problems.

(viii) They provide permanent and effective learning.

The importance of teaching aids can be summarised as under in the words of Edgar Dale—"Because audio-visual materials supply of concrete basis for conceptual thinking, they give rise to meaningful concepts—the words enriched by meaningful associations. Hence they offer the best anti-dote available for disease of verbalism".

Procedure of Selection

Following principles be kept in mind while making a selection of teaching aids for use in teaching a particular topic:

(i) The aid should have a relevance to the topic to be taught.

(ii) The aid must be such so as to suit the topic and helps to make the study of the topic interesting.

(iii) Any teaching aid used should not only be interesting and motivating but it also have some specific educational value.

(iv) The aid to be used should be a beat possible substitute in terms of reality, accuracy and truthful representation of object or the first hand experiences.

(v) The aid should be simple.

(vi) The aid should suit the physical, social and cultural environment of the pupils.

(vii) The teaching aid be easily available.

(viii) The teaching aid must help in proper realization of stipulated learning or instructional objectives of topics in hand.

The Utility

Teaching aids should be used properly to make teaching more effective. Teaching can because more effective if such aids are used widely but the use of such aids cannot provide a guarantee of good teaching. Following points are important for a proper use of teaching aids:

(i) Teaching aids should be woven with class-room teaching and these aids should be used only to supplement the oral and written work being done in the class.

(ii) While making use of any teaching aid an effort be made that the teaching aids being used in any class are in confirmity with the intellectual level of the students and is in accordance with the previous experience of the students.

(iii) Only such aids the preferred which provide a stimulus to the students for greater thinking and activity.

(iv) If possible actual specimens be preferred to a photograph or a slide of a specimen.

(v) The teaching aid used should be exact, accurate and real as far as practicable.

(vi) The Teacher should use a teaching aid only when he is quite sure about handling a specific teaching aid. For handling some aids (*e.g.* operating a projector etc.) training is provided by various authorities. For this purpose more information can be obtained from local SCERT or directly from NCERT, New Delhi.

(vii) Teaching aids used be such as are closely related to pupils experiences.

(viii) The teacher should use a teaching aid only after a proper planning so that the aid is used exactly at the point; in the process of teaching, where it best fits in the process of teaching.

(ix) Teacher should see that a follow up programme follows the lesson wherein a teaching aid has been used.

(x) Teacher should carry out occasional evaluation about the use, function and effect of a teaching aid on the learning process.

Different Types

For convenience of discussion the teaching aids may be grouped as under:

(i) Visual aids

(ii) Aural aids

(iii) Audio-visual aids

(iv) Activity aids and

(v) Memory aids

Aids of Various Kinds

Under this head we will take of following types of teaching aids:

(a) Displayboards such as Chalkboards or Blackboards, Flannelboards, Bulletinboards, Magneticboards etc.

(b) Charts, pictures and models.

Visual aids are those which can be appreciated and understood by seeing them only.

Displayboards. It is any flat surface that can be used to white information to be communicated. At present for this purpose the use is made of *blackboard* or *chalkboard, bulletinboard, flannelboard, magnetic-board* etc.

Though material for display on such a board can be collected from any source even from a text book but for being effective the material should be displayed in such a way that it is eye catching, colourful and purposeful.

Blackboard or Chalkboard. It is one of the most common visual aids in use. It is a slightly abrasive writing surface made of wood, ply, hardboard, cement, ground glass asbestos, state, plastic etc. with black, green or bluish green paint on it. Details of various types of chalkboards and their arrangement for a science laboratory have been given in the lessons dealing with these topics. A chalkboard is generally installed facing the class which is either built into the wall or fixed and framed on the wall and provided with a ledge to keep the chalk sticks and duster. Portable chalkboards are also available these days. Such chalk boards can be placed on a stand with adjustable height. Generally white chalk sticks are used for writing on the blackboard or chalkboard but some times coloured chalk sticks are also used. The coloured chalk sticks are used for better illustration.

Characteristics of a Good Chalkboard. Some of the characteristics of a good chalkboard are as follows:

(i) Its surface should be rough enough so that it is capable of holding the writing on the board.

(ii) Its surface should be dull so that it can eliminate glare.

(iii) Its surface should be such that the writing on the board can be easily removed by making use of a cloth or a foam duster.

(iv) Its height should be so adjusted that it is within the easy reach of the teacher and is easily visible to the students.

Effective Use of Chalkboard. We find that chalkboard is the most common teaching aid used by the teacher for writing important points, drawing illustrations, solving problems etc. The chemistry teacher should keep the following points in mind to use the chalkboard effectively.

(i) Write in a clear and legible handwriting the important points on the chalkboard but avoid over crowding of information on the chalkboard.

(ii) The size of the words written on blackboard should be such that they can be seen even by the back-benchers. The letters should not be less than one inch in height. The recommended height of letters on a chalkboard in between 6 cm to 8 cm. For this the teacher should frequently inspect his own chalkboard writing from the view point of the back-bench on a corner seat.

(iii) There should be proper arrangement of light in the class room so that the chalkboard remains glare free.

(iv) To emphasise some points or parts of a sketch or a diagram coloured chalks be used.

(v) Rub off the information already discussed in the class and noted down by the students.

(vi) Draw a difficult illustration before hand to save the class time.

(vii) Stand on one side of the chalkboard while explaining some points to the students.

(viii) Make use of a pointer for drawing attention to the written material on the chalkboard.

(ix) Students may be allowed to express their ideas on chalkboard, or to make alterations or corrections. Some times teacher may intentionally draw some incorrect diagram and ask the students to make necessary correction, alteration etc.

(x) For maintenance of proper discipline in the class the teacher should always keep an eye on his class while writing on the blackboard.

(xi) For proper writing on chalkboard the chalk stick be broken into two pieces and the broken end of the piece be used to start writing.

(xii) While writing on a chalkboard keep your fingers and wrist stiff and move your arm freely.

Advantages of Chalkboard. Some of the advantages of chalkboard over other visual aids are as follows:

(i) It is a very convenient teaching aid for group teaching.

(ii) It is quite economical and can be used again and again.

(iii) Its use is accompanied by the appropriate actions on the part of the teacher. The illustrations drawn on the blackboard captures students attention.

(iv) It is one of the most valuable supplementary teaching aid.

(v) It can be used as a good visual aid for drill and revision.

(vi) These boards cart be used for drawing enlarged illustrations from the text books.

(vii) It is a convenient aid for giving lesson notes to the students.

Limitations of the Chalkboard. Some of the important limitations of a chalkboard are as under:

(i) The use of chalkboard makes students very much dependent on the teacher.

(ii) It makes the lesson teacher paced.

(iii) It makes the lesson dull and of routine nature.

(iv) It gives no attention to the individual needs of the students.

(v) Due to constant use chalkboards become smooth and start glaring.

(vi) While using chalk-sticks to write on chalkboard the teacher spreads a lot of chalk powder which is inhaled by teacher and students and it may affect their health.

Bulletinboards. It is a display board on which learning material on some scientific topic is displayed. It is generally of the size of a blackboard but some times even bigger depending on the wall space available. It is generally in the form of a framed softboard or strawboard or corkboard or rubber sheets. Such bulletin boards can be specified for individual branches of chemistry or even for some specified chemistry topics *e.g.* chemistry puzzles, chemistry news, chemistry cartoons etc. such a board can also be used for displaying the best work of students. However for a all purpose bulletinboard the following type of display material in recommended:

(i) Interesting science news.

(ii) Book Jackets of recently published chemistry books.

(iii) Brochures.

(iv) Cartoons.

(v) Poems.

(vi) Sketches.

(vii) Pictures.

(viii) Photographs.

(ix) Thoughts.

(x) Announcements etc.

An effort be made to change the material on bulletinboard as frequently as in practicable. Whenever the teacher starts a new topic he may ask the students to display the concerned material on the bulletinboard and the teacher should specifically mention to the students the display material on the bulletinboard while teaching a topic to the class. Students be asked to take the charge of bulletinboard by rotation'.

How to Use a Bulletinboard. To make use of bulletinboard as a useful teaching aid the bulletin board be used for creating interest amongst students an specific topics. For effective use of bulletinboard as a teaching aid following points be kept in mind:

(i) Effort be made jointly by the teacher and the students to procure material from various sources on a given subject or topic.

(ii) Before displaying the material on the board sort out the material relevant to a specific subject or topic.

(iii) Make best use of your aesthetic sense to display the material on the bulletinboard.

(iv) Do fix a title for the specific subject/topic of display material on the top centre of the bulletinboard.

(v) It is desirable if a brief description about the specific subject or topic is fixed below to title.

(vi) The height of bulletinboard from ground level be about 1 m.

(vii) The bulletin board be fixed in an area where enough lighting can be provided.

(viii) The material displayed should be large enough and should be provided with suitable headings.

(ix) Over crowding of material on bulletinboard be avoided.

Advantages of Bulletinboards. Some of the advantages of bulletinboard as a teaching aid are as follows:

(i) It is a good supplement to class room teaching.

(ii) It helps in arousing the interest of students in a specific subject/topic.

(iii) It can be effectively used as a follow up of chalkboard.

(iv) Such boards add colour and liveliness and thus also have decorative value in addition to their educational value.

(v) Such boards can be conveniently used for introducing a topic and for its review as well.

Limitations of Bulletinboard. Some limitations in the use of bulletinboards as teaching aids are as follows:

(i) They cannot be used for all inclusive teaching.

(ii) They can be used only as supplementary aids to some other teaching aid.

(iii) At times it becomes very difficult to make proper selection of the display material for certain topic.

Flannelboard. It is also some times referred to as *flannel graph* or *felt board*. It is made of wood, cardboard or strawboard covered with coloured flannel or woolen cloth. It is one of the latest devices effectively used for science teaching. Display materials like cut-outs, pictures, drawings and light objects backed with rough surfaces like sand paper strips, flannel strips etc. will stick to flannel-board temporarily.

For display purposes a flannelboard of 1.5 x 1.5 m is generally used. It can be fixed next to the blackboard or can be placed on a stand about one metre above the ground.

How to use a Flannelboard. Following points be kept in mind for effective use of flannel board as a teaching aid:

(i) The teacher should collect a large number of pictures or wall cut diagrams etc. and back them with sand paper pieces. He may then make use of these by displaying there on the board one by one, after proper selection.

(ii) Display the material on the flannel board in a sequence to develop the lesson.

(iii) Make proper use of flannel board for creating proper scenes and designs relevant to the lesson.

(iv) Change the display material on the board as frequently as required.

(v) Flannel board can be used quite effectively for showing relationship between different parts or steps of a process.

Advantages of Flannelboard. Some of the advantages of using flannelboard as a teaching aid are as follows:

(i) It is quite economical and easy to handle and operate.

(ii) The pictures or cuttings can be easily fixed and removed when required, without spoiling the material. Thus same material can be used for display many a times.

(iii) Any display material on the board holds the interest of students and arrests their attention.

(iv) Such boards enable a teacher to talk along with changing illustrations to develop a lesson.

Magnetic Chalkboard. It is a framed iron sheet having porcelain coating in black or green colour. Such a board can be used either to write with chalk sticks, glass marking pencils and crayons or to display pictures, cut-outs and light objects with disc magnets or magnetic holders.

Thus such a board functions both as a chalkboard and as a flannelboard. We can display visual learning material on such a board while writing key points on it. Such a board provides the flexibility of movement of visual material. It is possible to display even a three dimensional object on such a board using magnetic holders.

Since the magnetic chalk-board functions both as a chalkboard and as a flannelboard so various points discussed for the effective use of these boards be kept in mind while using magnetic chalkboard as an effective teaching aid.

Advantages of Magnetic Chalkboard. Some of the advantages of magnetic chalkboard are as follows:

(i) It is a versatile teaching aid that combines the advantages of both a chalkboard and a flannelboard.

(ii) It is possible to move visual material by sliding it along the surface of the board such a movement is not possible on a flannelboard.

(iii) It is very light and can be easily taken from one place to another.

(iv) Such a board can be easily got prepared in the school from an iron sheet and painting with some good paint.

Charts, Pictures and Models. Charts, pictures and models also are an important teaching aids.

Charts. Some times charts are needed by the teacher to supplement his actual teaching. There are certain charts where in the interior of some thing is depicted *e.g.* various systems of human body, internal combustion engine, motor car etc.

Following points be kept in view while using charts as teaching aids:

(i) An effort be made to use charts prepared by students under the guidance of the teacher, however some charts may be purchased.

(ii) Duly such charts be purchased which have bold lines and in which such colours are used as could be seen and distinguished even by the back-benchers.

(iii) Charts should give only the essential details.

(iv) Charts should be properly and clearly labelled in block letters.

Sources for Procurement of Charts

(i) Charts can be prepared by students and teacher.

(ii) Charts can be purchased.

(iii) Charts can be procured on a very normal cost from the following sources:

(a) Ministry of Education, Govt. of India, Delhi.

(b) NCERT, New Delhi.

(c) Director, Extension Service of College of Education in the State.

(d) SCERT of the state.

(e) District Public Relation Officer.

Advantages of Charts

(i) They can be made quickly.

(ii) They have a better appeal.

(iii) Only bare essentials can be shown in the chart and unnecessary details can be avoided.

(iv) Charts are available from various sources.

Pictures. Pictures of gas-works, steamships, and locomotives and portraits of great men of chemistry—chemists will be of great help in teaching of chemistry provided a reference in made to them. Portraits of great scientists if displayed in chemistry room give it the proper scientific atmosphere. These pictures, portraits etc. can be used as teaching aids and they are quite useful in a demonstration lesson. Everything a child learns can be presented graphically with the aid of pictures and brightly coloured diagrams which will excite his interest.

Following points be given due consideration while using pictures as teaching aids:

(i) Pictures should be bold, direct and sufficiently large.

(ii) Pictures should not be over loaded with information rather they should stick to the maxim, *one picture, one idea*.

Models. In teaching of science models are very frequently used. Various costly models are available and some of these may be available and in school laboratory. However the cost of such models should not be any hindrance to the use of models as teaching aid because a science teacher can prepare almost all types of models by making use of ingenuity. It is also possible to take some very costly models on loan or such models can even be hired. Models are very helpful in making the subject clear to the students and they also give the student an idea of the actual shape/size etc. of the article under discussion.

In using charts, pictures and models as teaching aids the teacher should be careful to plan their proper display. These should be displayed in such a way and at such a height that each student can have a detailed view of it.

Following is the list of some firms from whom scientific charts and models can be procured:

1. M/s Scientific Instruments Stores, J-355, New Rajinder Nagar, New Delhi.
2. M/s Educational Aids and Charts, 20, I Block, Kumara Park, West Extension, Bangalore-20.
3. M/s Variety Teaching Aids, Bagalkot, Distt. Bijapur.
4. M/s Educational Emporium, 15-A, Chittranjan Avenue, Calcutta-7.
5. M/s Oxford University Press, Appollo Bunder, Bombay.
6. M/s School Aids Manufacturing Co., 12-Gun Boat Street, Fort, Bombay-1.
7. The Director, Survey of India, Hathi Barkala Road, Dehradun (UP).
8. M/s Hobby Centre, Mount Road, Madras-2.

Aural Aids: In this type the following aids are considered:

(i) Broadcast talks,

(ii) Gramophone lectures, and

(iii) Tape recordings.

Broadcast Talks. All India Radio has in its regular feature some programmes meant for school children. In such a programme generally talks on educational matters or on scientific topics are broadcasted. Such a talk in quite useful for students as also for chemistry teacher. The topic, date and time of broadcast of such talks are given an advance by All India Radio. A school can take benefit of such talks only if it possesses a good radio set and a period is provided in the school time-table for listening such talks. Such an arrangement can be worked out by the school authorities and then teacher can refer to such talks while teaching his class. It is also possible to synchronise the broad cast talk an some topic with the actual teaching of that topic in a class.

Some handicaps of such broadcast task are listed here:

(i) Sometimes when the receiving set is not working satisfactorily there prevails a sense of strain in the class room.

(ii) Some students are poor listeners and may not be benefited by such talks although they benefit by normal teaching through questions, demonstrations and reading.

For the maximum utility of such talks following points be kept in view:

(i) The students with bad hearing be seated on front seats.

(ii) To keep students interest alive in such talks teacher should tell his students in advance a few questions which they have to answer after the talk.

(iii) Only short duration talks be arranged.

Such talks cannot be a substitute to the actual teaching and such a talk is only to help in teaching.

Gramophone Lectures and Tape Recordings. Another teaching aid available to a science teacher is records of short talks an interesting scientific topics by eminent scientists, doctors etc. Magnetic tapes of such recorded talks are now available and the talk can be easily reproduced in the class room. These talks provide an inspiration to the students and such a talk once recorded can be used again and again. Such recording can either be used to introduce a topic or to develop a topic.

Audio-Visual Aids: In this category those teaching aids are included which involve the use of two of our senses *i.e.* hearing and seeing. These are classified as (i) optical aids and (ii) Television.

Need for A.V. in Teaching

Audio-visual aids are very important in teaching of chemistry because of the following reasons:

(i) Sensory experience is the foundation of intellectual activity. Verbal symbol, which is meaningless becomes meaningful when it is associated with visual symbols. For example meaning of precipitate is understood only when it is seen in test-tube.

(ii) A.V. aids are needed to stress facts and concepts in chemistry teaching.

(iii) Mental growth is the outcome of two antithetical processes i.e. differentiation and integration. Differentiation develops out of integration. Audio-visual aids are more useful in process of differentiation.

(iv) Generalisation attains a meaning and it becomes concrete experience only with the help of A. V. Aids.

(v) A. V. aids also help in increasing the vocabulary of pupils.

Optical Aids: Some such aids are discussed here. *Magic Lantern (or Glass slide projector).*

Psychologists have now confirmed that a child grasps abstract facts slowly and can only remember a name which recalls some definite reality. Thus he should be confronted with visual teaching aids to broaden his experience.

A *magic lantern* is a simple device used to project pictures from a glass slide on a screen or wall. Teacher can make use of this device when he intends to show some small figure or illustration to whole class. Many a schools have a *magic lantern* in their laboratories as it is not very costly slides are readily available in the market on various chemistry topics. These can also be got prepared on demand and the cost of such a slide is quite reasonable. Such slides can even be prepared by science teacher himself after some practical training which can be provided by extension service department of training colleges. colleges. Epidiascope

Epidiascope is a more costly instrument but it can project opaque objects as well as transparent objects. The pictures projected by epidiascope are much brighten and need, a less powerful light so that room need not be absolutely dark. Epidiascope can be used to project any picture, map, diagram, photograph or small object. No slide is needed for projection with an epidiascope.

The name *epidiascope* is given to this machine because of the tact that it works, as an *episcope* when it is used to know the image of an opaque object. This machine can be used to project slides and this is possible just by moving a lower provided for the purpose. When it is used to project a slide then at serves as a *diascope*. Thus *epidiascope* is a combination of these two *i.e. episcope* and *diascope*.

Advantages of Epidiascope. In comparison to other projection machines *epidiascope* has some advantages. Some of these are as follows:

(i) It can be operated in a room which may not be absolutely dark.

(ii) With the help of this machine original colours of the picture or photograph can be projected.

(iii) The projection on the screen can be kept for some time during which teacher can explain and discuss it in the class.

(iv) It provides teacher an option to handle the lesson according to himself.

Following points provide useful hints for the proper handling of an epidiascope:

(i) The apparatus works well in a dark room.

(ii) While projecting with an epidiascope an effort be made to keep exposed to the head of the lamp for minimum time delicate pictures, photographs or other such objects.

(iii) The person handling the apparatus must be given some practical training before he is allowed to handle the machine.

Film Projector, Micro-Projector, Film-Strip Projector

There are further improvements on the teaching aids discussed so far. These have brought about a revolution in teaching of science. Science films are shown to the students to illustrate various applications and uses of science as also to supplement the class room teaching. Both type of films have some basic objectives to serve.

Film Strip-Projector

It is an improvement on magic lantern and this machine can be used to project many a topics on a single strip. One such strip generally consists of 40-100 separate pictures and such film strips are available on loan from Central Film Library, NCERT, New Delhi. On such a film strip pictures concerning one topic are arranged in a definite order.

This machine can be easily handled by the chemistry teacher. The machine is operated by hand and thus can be stopped at the discretion of the teacher whenever he wants to explain some aspect of a topic being shown on machine.

Micro-Projector

This is less commonly used in chemistry teaching. This projector is generally operated in a dark room. The projection can be taken on vertical screen if whole class is expected to see it. However such a film cannot be distinctly seen by a student if he is sitting at a distance more than 12 feet from the screen.

Film Projector

This machine is used for showing chemistry films. Some good science films on various topics are available and there can be had an loan some times even free of charge from the source, given below:

(i) Central Film Library, NCERT, New Delhi.

(ii) U.S. Information Service, New Delhi.

(iii) British High Commission Office, New Delhi.

(iv) Some Other Embassies, New Delhi.

For projecting this films in school generally 16 mm projector ('RCA', 'Bell and Havell') are used. These 16 mm projectors are less costly and easier to transport as compared to a 35 mm projector.

Advantages of Motion Pictures. There are some definite advantages of motion pictures to be used as teaching aids, some of these are as follows:

(i) They draw attention of the students.

(ii) They help to bring past to the class-room.

(iii) It is possible to reduce or enlarge the size of the object by using the machine.

(iv) They can be used to show a process which a naked human eye cannot see without its aid.

(v) They can be used to show a record of an event.

(vi) They can serve a large class at a time.

(vii) They provide a good aesthetic experience.

(viii) They help in understanding relationship between things, ideas and events.

Precautions. The teacher should take the following precautions whenever he wants to use a film projection as a teaching aid:

(i) He should satisfy himself about the lighting management and seating arrangement in the room where such a film show is to be given.

(ii) He should himself see the film before hand.

(iii) He should give a complete background of the film to the students before the actual screening of the film.

(iv) He should see that complete calm and peace is maintained during the screening of the film.

(v) Immediately after the film show, he should invite comments, questions etc. from the students and try to answer all the quarries of the students.

(vi) He should encourage some of his students to write articles etc. based on the film show and such articles etc., may be shown on well magazine, may be printed in school magazine.

Television : The role of television in the present day world is becoming more and more important and it is one of the most important teaching aids. It combines the advantages of a radio (broadcast) and of a film. This can be used for mass education and now U.G.C. programmes are a regular feature on "Door Darshan". The topics of discussion are announced in advance and lesson from well qualified reasons and specialists in their fields are shown on TV. Teacher can easily plan his work accordingly and in this way he can make use of TV as a teaching aid.

Limitations : The use of A.V. aids in teaching of chemistry has the following limitations:

(i) The use of A.V. aids is not a guarantee of successful teaching.

(ii) A.V. aids are not a clear substitute for oral or written methods of gaining knowledge.

(iii) Visual instructions are sometimes confused with entertainment.

(iv) Visual aids vary in their effectiveness in direct proportion of their degree of reality.

11

Laboratory for Practicals

Chemistry is essentially a practical oriented subject. No course in chemistry can be considered as complete without including some practical work in it. For proper understanding of chemistry, it should be taught using a large number of demonstration experiments. For carrying out demonstration experiments and for the performance of practicals by the students, a chemistry laboratory is a must for every school offering chemistry as a subject. Like any other science subject a chemistry laboratory is justified on the following grounds:

(i) In chemistry laboratory the required apparatus and other apparatus etc. can be safely stored.

(ii) As in other science subjects so also in case of chemistry laboratories are helpful in creating and promoting scientific attitude in the pupils.

(iii) Laboratory provides a proper and congenial place for performing experiments and is helpful in developing a sense of cooperation and spirit of healthy competition among the students.

Value of Practicals

No course in chemistry can be considered as complete without including some practical work in it. The practical work is to be carried out by individual in a chemistry laboratory. Most of the achievements of modern chemistry are due to the application of the experimental method. At school stage practical work is even more important because of the fact that we 'learn by doing' scientific principles and applications are thus rendered more meaningful. It is a well known fact that an object handled impresses itself more firmly on the mind than an object merely seen from a distance or in an illustrations. Centuries of purely deductive work did not produce the same utilitarian results as a few decades of experimental work. Practical class room experiments help in broadening pupil's experience and develop initiative, resourcefulness and cooperation. Because of the reasons discussed above practical work forms a prominent feature in any chemistry course.

Practicals Done

Out of the various teaching methods discussed earlier *the Assignment method* is the only method that combines theory and practice in a harmonious manner and can be easily practiced in our schools. The *Heuristic method* is predominently a *laboratory method.* However from this it should not be concluded that practical work in laboratory is impossible if the teacher makes use of any other teaching method. Thus irrespective of the method adopted by the teacher for teaching of chemistry in the class, practical work in laboratory must be attempted. The following guide lines will help the chemistry teacher to make his practical work effective.

Basic Rules

For smooth working in the laboratory teacher should give due consideration to the following points:

(i) If teacher follows the demonstration method to teach theory, he should remember the most important principle that practical work should go hand in hand with the theoretical work. Thus if a class is doing theoretical work in chemistry it should also do practical work in chemistry during the practical periods.

(ii) An attempt be made to arrange the practical work in such a way that each student is able to do his practical individually. Thus for practical work individual working be preferred in comparison to working in groups.

(iii) In case of a large class, it is convenient to divide the class in a suitable number of smaller groups, for practical work. A practical group in no case should have more than 20 students. The limit on practical group is essential otherwise teacher will not be able to devote individual attention to the students.

(iv) To save time on delivering a lecture about do's and don'ts in laboratory, card system is used. This card which contains certain amount of guidance printed on it is given to each pupil. In some laboratories where card system exists each student is given a card containing instructions about the experiment that he has to perform. This card also contains the details of the apparatus required. Student can complete his practical work according to instructions given in the card.

(v) The apparatus provided should be good so that students get an accurate result particularly in those experiments in which the student is likely to compare the numerical value of his result with some standard. However, every chemistry teacher should gaurd against 'Cooking' of results by his pupil. If this bad habit of cooking is not checked in the beginning it persists through out the students career.

(vi) A true and faithful record of each and every experiment be kept by pupils. The record should be complete in all respects.

(vii) To check the habit of 'cooking' teacher should see that students enter all their observations directly in their practical notebook. The teacher should insist that the pupils do not go to the balance room without first entering the data in their notebooks.

(viii) Students should not be allowed to erase any figures. To change any wrong entry the same be crossed and correct figure entered only with the permission of the teacher.

(ix) Students should not be allowed to calculate results or write data on scrap papers.

(x) In practical notebook the right hand page be reserved for record while the left hand page be left for diagram and calculations. This practice be followed for Assignment method. For any other method the laboratory work be done on left hand page of practical note book and procedure etc. on right hand page of practical note book.

(xi) Teacher should see that students complete their practical note book in all respects and get it signed before they are allowed to leave the laboratory. Incomplete practical note books be kept in the laboratory and students be asked to complete it in their spare time.

(xii) Teacher should thoroughly check and critically examine the account written by students.

(xiii) Whenever a student is required to make use of a piece of apparatus for the first time it is the duty of the teacher to explain to his students the working of the apparatus. He should also explains reasons for necessary care and accuracy.

(xiv) Teacher should see that students find no difficulty to get apparatus and chemicals needed by them. In the absence of provision for laboratory assistants it is for the teacher that he arranges the apparatus in such a way that things frequently needed by students are easily accessible to them. Teacher should also emphasise proper and economical use of apparatus and chemicals.

(xv) While working with larger groups and with limited apparatus teacher can act as under:

(a) He may use assignment method.

(b) He may allow students to work in groups.

(c) He may devise alternate simple experiments and work with improvised apparatus.

(d) He may allow use of home made apparatus.

(xvi) Whenever the teacher is required to draw up suitable laboratory directions or instructions for practical work by pupils, be should keep the following points in mind:

(a) Beginner be given detailed directions.

(b) He should not tell the students what is actually going to happen.

(c) The main aim of the experiment should be made clear.

(xvii) During a practical class teacher should observe all children from his desk otherwise chances of accidents are there.

Even when teacher has to move from his desk his power of control over the class should be such that students continue their work satisfactorily.

Combined Lecture room-cum-Laboratory : Laboratory is a spacious room where in a group of students carry out their practicals. The work of designing and building a science room (laboratory and lecture room) is that of the architect but science master should collaborate with the architect in planning for what is best from the educational point of view. The plan of a combined lecture room and laboratory for use in schools upto matriculation standard, devised by Dr. R.H. Whitehouse, formerly principal of the Central Training College, Lahore, has been adopted as the official standard plan by Punjab Education Department.

The plan combines laboratory and class room for science teaching. The suggested size of the room is 45′ × 25′ and it is meant for a class of 40 students which is sub-divided in two groups of 20 each for practical work.

The size of the room is most economical. Though the length of the room is 45′ but it should not be considered as disadvantageous because the teacher is expected to address a class of 40 students who will be occupying only about half the room.

For constructing such a room walls are to be of 1′ 6" thick keeping Indian conditions in view, use of distemper be preferred to white wash for the walls. A perfectly smooth floor is preferable to one exhibiting any roughness. Such a floor is easier to clean of the two doors, one is used for lecture room and the other is reserved for laboratory part. To provide side lighting three large windows (6′ × 8′) are provided. One of these is provided near practical benches and two near seating accommodation. Doors as also windows sills may be used as shelves for carrying out experiments. To avoid flies wire gauze screens be provided to the windows. If necessary, in such a case, the windows be constructed with an upper and a lower half. The lower half is fixed so that the inner sills of windows could still be used as shelves.

The Equipments

In the area meant for lecture room a wall black board 10′ × 4′ is provided. About 3′ away from this black board is the teachers table which is about 6′ long and 2.5 feet high. Such a table can be conveniently used both as a writing table as also a demonstration table and causes no disturbance or in convenience to the students in watching the demonstration or observing the black board.

For seating dual table and chairs are most economical. Thus by providing twenty tables and forty chairs sufficient seating arrangement could be made. Dual tables should be of the size 3.5′ × 1.5′ × 2′. They may be provided with shelf. The top of these tables should be flat and plain having grooves for pen/pencils. The chairs are 1.5′ high in the seat, which in case of an iron chair, may be covered with a small mat. The area necessary for a dual table and two chairs is a square of 3.5′. Passages of 1.5′ are sufficient for single file and 2.5′ to 3.5′ at the sides.

A sink is provided for use of the teacher. The size of the sink generally used is 18″ × 12" × 6".

The advantages of table and chair system are as under:

(i) They are quite economical.

(ii) They provide quite natural seats.

(iii) They allow enough space for easy passage of the students.

(iv) They can be easily moved while cleaning the room.

(v) They can be used for other purposes such as accommodating guests at various school functions.

Laboratory : In the laboratory part of the room are provided six laboratory tables which are made of wood and are perfectly plain. A black board is also provided on this side of the room. The laboratory tables are of the size 6′ × 3.5′ and are provided with a shelf on the working side just below the top. Four students can work on each table. The whole of each table except top should be stained dark. The top should be treated with wax ironed with a hot flat iron in order to fill the pores of the wood and to prevent the easy penetration of the liquids. The space between 3′ and 4′ and passage-way at the end of the tables is 2′ wide. At school level the laboratory tables are not provided with any sink. Some of the reasons for not providing the sinks are as follows:

Economy: A large economy is observed because much plumbing and a network of drains is avoided. Cost of sinks is also saved. For most of the experiments at school level a trough can serve the purpose.

Usefulness: The table is quite useful for both physics and chemistry. In absence of sinks more space is available for use as working space. Such a table can also be used for other purposes.

Appearance and cleanliness: The floor of the room is not broken for providing drains etc. It gives a better look.

Tidiness: The tables if provided with sink would make the room untidy because such tables invariably allow splashing of water which is likely to interfere with experiments and is likely to create problems.

As shown in the plan there are only three sinks, one for the teacher and two for the students. Of the two sinks for students one is placed in the window recess and the other in recess in the wall. Each of the sinks is provided with a drawing board having grooves arranged to drip over the suits. It is used for placing beakers, flasks, etc. for drying.

For placing balances, recess in walls may be used. They may be about a foot wide at a height of about 3′, 3″. Such recess has the following advantages over wooden or stone shelf:

(i) It is very economical because only very small masonry is needed.

(ii) It is more substantial as compared to a bracket shelf.

(iii) It does not project into the room and so space economy can be made.

For providing ample accommodation for balances a length of 7′ to 7.5′ is sufficient.

In the plan provision has also been make for the storage of apparatus, equipments etc. For this purpose there is a provision of eight almiras (each with 7′ × 5′ dimensions). Each almirahs provided with shelves 1.5′ deep, of this 1′ is recessed in the wall and only 6″ projects out. These almirahs provided sufficient space for the storage of not only the apparatus, equipment etc. but can also serve the purpose of storage of science library.

Reagent shelves can be very conveniently placed on either side of the recesses for balances space can also be found, for placing notice boards for assignments of work, results of tests, etc., on the wall between the windows or just inside the doors.

The Advantages

The combined lecture room-cum-laboratory discussed in previous pages has the following advantages:

(i) It is very economical.

(ii) It is compact and provides enough space for seating, working, storage etc.

(iii) It can be furnished easily and with meager resources.

(iv) It provides enough and comfortable seating space for the students.

(v) In this room science atmosphere prevails.

(vi) It provides an opportunity for better control. For a better control following points be kept in view by a teacher:

(a) Every student has his assigned place which is indicated by his name written on a card placed in a brass card holder fixed on the leg of the table.

(b) The four boys working on any table be allotted number 1, 2, 3, 4 and number 1 of each table be asked to collect four sets of articles required for each table. Number 2 be asked to remove the dirty apparatus, after the period, to drain board and number 3 will remove clean apparatus. Number 4 will wipe down the table with a duster.

(c) Class monitors be named for cleaning dirty apparatus after school hours or during recess period.

(d) Students be made responsible for the correct alignment of their tables. For this black and white lines be pointed on the floor.

The Planning

In a senior secondary school the arrangements are made to provide education in chemistry as elective subjects in addition to teaching of general science. In senior secondary school a provision has to be made for a chemistry laboratory. The laboratory in senior secondary schools is almost the same as in colleges. Each laboratory is provided with a preparation-cum-store room attached to it. The size of the laboratory will depend on the number of students likely to work in it at a time. About 30 sq. feet. If space be provided for each student. The structural details are generally provided by the architects but the following points be kept in mind.

Planning : It would be better if chemistry teacher is consulted and for this there should be frequent conferences between the chemistry teacher and the architect. Various points be thoroughly discussed. Some of the points of consideration are as under:

(i) Laboratories and class rooms should not be mixed on the same corridor.

(ii) Laboratories be situated, as far as possible, away from crafts room, music room, play fields, main gate etc.

(iii) The consideration be given to proximity of stores, preparation room, balance room, green houses etc.

Following points be given due consideration while planning individual laboratories:

(i) Each student is easily accessible to the teacher.

(ii) There is minimum of movement.

(iii) Each student has a cupboard, bottles, heating point and a sink near him.

(iv) Teacher can easily watch each student.

(v) Black board is visible to each student.

(vi) Each student can easily see the demonstration.

(vii) There is enough space between two laboratory tables.

(viii) Master switches be provided to control electricity, gas, water etc. in each laboratory.

Lighting : Proper lighting arrangements be made for laboratory tables and class rooms. Special attention be given to the lighting of demonstration table and black board. It would be preferred if a provision could be made for electrical lights over tables through pulleys so that their height may be varied from 2 to 8 ft. Two way switches be provided for controlling the main lighting from doors and preparation rooms. Dark blinds or curtain must be provided for each laboratory.

Ventilation : It possible each laboratory should be surrounded by a 6′ verandah on all sides to keep away the direct heat of the sun. Ventilators be provided as usual. In case of chemistry laboratory ceiling should be high and exhaust fans must be provided.

For laboratory ventilation full height windows are desirable. To avoid direct draughts even when the windows are open windows should be designed in Fig.

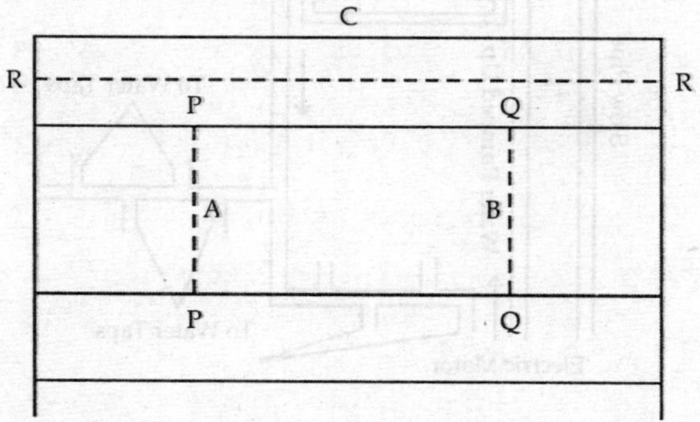

Fig. Design of a window to avoid draughts.

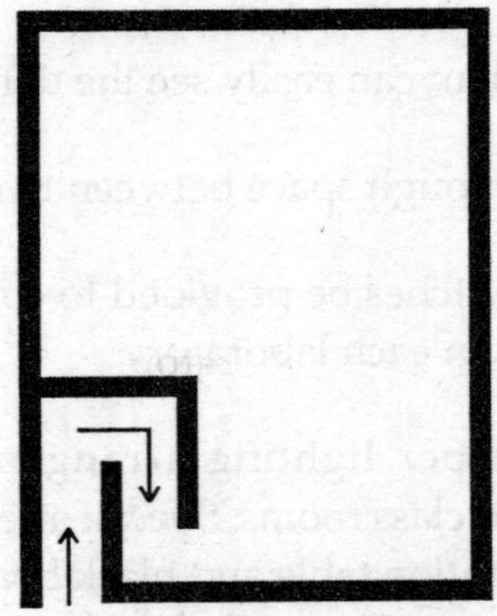

Fig. A 'no-door" Dark Room

Windows A and B are side lung or pivoted along the axis PP and QQ. Window C is pivoted along RR. If this plan is used them windows can open in different ways depending on the wind direction. In dark room the ventilation can be provided by deflecting incoming and outgoing air by opaque partition. A no door arrangement as shown in Fig. may made.

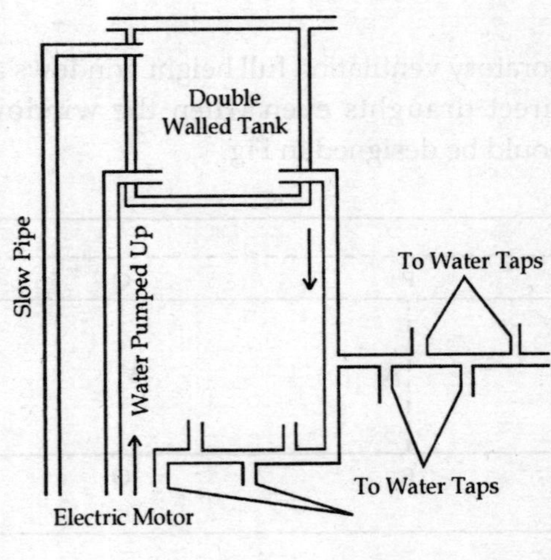

Fig.

Water Supply : Provision of water supply must be made in every laboratory. Water supply is most essential item and for this purpose proper arrangement of water laps and sinks is a must in every laboratory. In case of non-availability of adequate water supply from municipal/local sources alternate arrangements have to be made. For making alternate arrangements suggestion given below be considered.

A water storage tank having a capacity of 1000 to 5000 litres be constructed with concrete and cement or a readymade tank of synthetic material be purchased and such a tank be then placed at the roof of the room. Water be then lifted using electric pump for filling this tank. The water supply is then provided from this storage tank to the laboratories. A tentative scheme for storage and supply of water is shown in Fig.

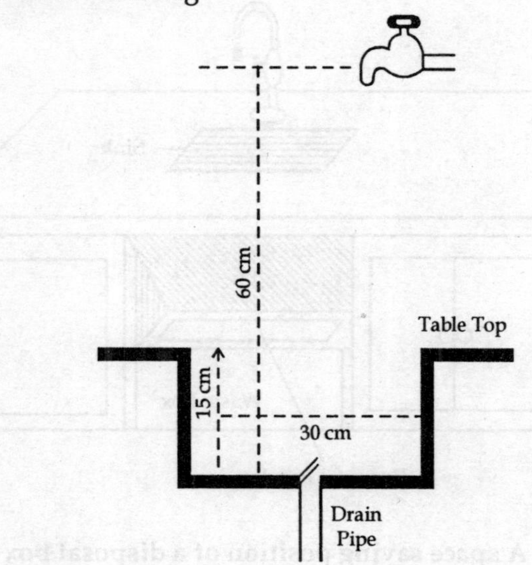

Sinks : Provision of sinks in each laboratory is one of the essential requirements. For a laboratory of ordinary size generally four sinks of 15" × 12" × 8" or 20" × 15" × 10" are sufficient. These sinks be fitted on side walls. These sinks are in addition to the one provided with the demonstration table. Waste water from these sinks is carried to the drains with the helps of the lead pipes fitted with the sinks. In laboratories kitchen type sinks are preferred to wash basis type. Fig gives a sketch of sinks and drainage for water.

Waste Disposal: In laboratories two types of wastes (*i.e.* liquid and solid) are often encountered. Arrangements have to be made for disposal of these wastes. For disposal of liquid wastes use of lead pipes or earthen ware pipes is considered most suitable. However care be taken to avoid the flow of solids like pieces of filter paper, cork, broken glass pieces etc. through these pipes, otherwise these pipes get chocked. For disposal of such solid wastes metal boxes or wooden boxes be provided. Such boxes be placed in the corners of the laboratory and students be asked to put all solid wastes in these boxes. Such waste boxes can even be placed under the sinks as shown in Fig.

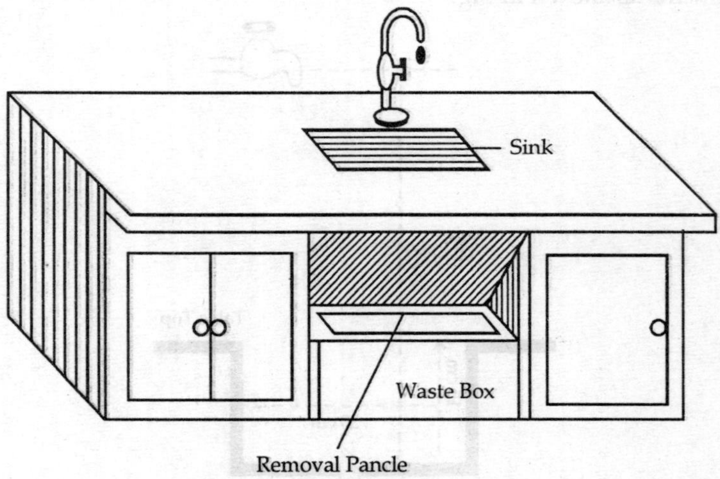

Fig. A space saving position of a disposal box

Pipe Lines: Installation of water pipes and gas pipes is another important aspect for furnishing a science laboratory. While installing pipes some of the points that be given due consideration are given as follows:

(i) Not more than 4 or 5 half inches pipes be led from any pipe for purpose of supply of water or drainage of water.

(ii) In case of physics laboratory all efforts be made to avoid iron pipes.

(iii) Pipes should never be placed on the laboratory tables.

(iv) It is convenient if the pipe fittings are not underground.

Gas Supply: For adequate supply of fuel gas to the laboratory generally any one of the following arrangements is made:

(i) Kerosene oil gas plant is installed.

(ii) Coal gas plant is installed.

(iii) Petrol gas plant is installed.

(iv) Gobar gas plant is installed.

The petrol gas plant is preferred as such a plant is economical and such plants are available in various capacities. A moderate capacity plant can feed 10-20 gas taps. Petrol gas plants are readily available and such plants are also manufactured at Ambala (Haryana). These plants can be easily operated.

For housing a petrol gas plant we need only a small room. The gas can be distributed to the practical tables using a 2" main gas pipe with further distributaries of 1/2" pipe. Each practical table is provided with gas taps and these taps should be of the rigid nozzle type and be fitted towards the back of the table. The gas tables be fitted in such a way that the point upwards and are at an angle of 45° from each other. If double benches are provided them taps should be fitted along the centre line of each bench. In most of the laboratories iron pipes are used but it would be preferable, in case of physics laboratory, if we use brass pipes. For controlling the supply of gas in addition to main control valve provision be made to control the supply of gas to each group of tables. These controls should be easily accessible to teacher and should not be easily accessible to students.

Laboratory Tables : The provision of laboratory tables as a must for each laboratory. The taps of laboratory tables be preferably made to teak wood. However other hard wood such as sheesham or deodar can also be used for making tops of laboratory tables. These tops are generally 1" thick. Other parts of the table *i.e.* legs, drawers etc. may be made of any other type of locally available wood. Plywood or hardboard can also be used for drawers. In chemistry laboratory such drawers are provided with the laboratory tables.

In addition to these provision for boards be made in the laboratories. For this either wall blackboards be provided or movable wooden blackboards with stands can be used.

The organisation of laboratories in secondary schools was also discussed at a seminar (All India) on the teaching of science in secondary schools. This seminar was held at Tara Devi (Simla) and it made some recommendations. Its recommendations are given below:

Layout

(a) Provision be made for one laboratory for chemistry in every higher secondary school.

(b) A floor space of 30 sq ft per student be provided in each laboratory.

(c) Adjacent store room be provided with each laboratory.

(d) A part of store room may be earmarked for use as a preparation room.

(e) In science wing, some suitable place for work benches with tools, be provided.

(f) A minimum of two class rooms provided with galleried seats be provided in each school.

The equipment for chemistry laboratory as recommended at the Tara Devi (Simla) Seminar is as under:

(i) Almirahs (wooden and steel).

(ii) Wallboard or blackboard.

(iii) Demonstration table (8′ × 4′) with cupboards, water and gas points.

(iv) Working tables with cupboards, shelves, water and gas points.

(v) Balance room should be attached to the laboratory.

(vi) Sinks on each working table or at least two large sinks at the corners of the laboratory.

(vii) A fume cupboard.

(viii) A wooden box half filled with sand for use as waste material box.

(ix) Acid proof drainage system.

(x) Shelves for reagent bottle on each working table and wall shelves for storage of reagent bottles.

The Guidelines

Instructions to Pupils : When a group comes to laboratory for practical work it should be given a guidance for the experiment to be performed. Such guidance can be given by (i) Laboratory instructions *(ii)* Laboratory manuals or *(iii)* Instruction cards.

Laboratory Instructions : Laboratory instructions should be given in a clear and simple language. It should give a precise but systematic method of performing the experiment. Special emphasis be given on how the record is to be kept and number of observations to be taken. Calculations be clearly explained and precautions to be observed be specifically emphasised.

Laboratory Manuals : Any good laboratory manual should contain the following:

(a) It should contain full and complete directions about the experiment. Such directions should clearly give the procedure to be followed and the precautions to be observed.

(b) It should give the method of recording of observations. Tables if needed for recording observations be clearly given.

(c) It should give clear directions about the writing and completing of practical note book. Important equations, diagrams etc, be given in it.

Instruction Card : In some laboratories instruction cards are used for providing guidance for laboratory work. Each student is given a card containing instructions about the experiment he has to perform.

The use of cards saves time as also wrong of the teacher. By this system different students can perform different experiments but the system is stereo typed and makes no allowance for individuals.

Usually cards of 6" × 4" are used.

Practical Records

For keeping a true and faithful record of practical work done by him each pupil is required to have a practical note-book. An effort be made to avoid printed note-books and plain note-books be encouraged. By using plain note-books teacher can make his students to follow any suitable method of keeping record of the experiment.

As a general practice description is given on the right hand page and observations are recorded on the left hand page. In case assignment method is used them the left hand page is reserved for preparatory work and laboratory record is kept on right hand page which contains description as also the observations.

The record of the method should be brief and in the first person singular. A three column system be used for keeping a record of experiments about the properties of gases etc. The three column be headed *Experiment, Observation* and *Inference*.

The precautions that have been observed while performing the experiment be recorded at the end in the first person singular (past tense).

All records in practical note-book be made with lead-pencil. The diagrams should be simple and will labelled.

Useful Items

Burns : Burns by Dry Heat (*i.e. by* flame, hot objects etc.). For slight burns apply *Burnol* and *Sarson oil*.

In case of blisters caused by burns apply *Burnol* at once and rush to dispensary.

Caution. Heat bums should never be washed.

Acid Burns : Wash with water and then with a saturated solution of sodium bicarbonate and finally with water. Even after this if burning persists, wipe the skin dry with cotton wool and apply *Sarson oil* and *Burnol.*

Caution. In case of conc, sulphuric acid, wipe it from the skin before giving the above treatment.

Alkali Burns : Wash with water and then with 1% acetic acid and finally with water. Dry the skin and apply *Burnol.*

Cuts : In case of a minor cut allow it to bleed for a few seconds and remove the glass piece if any. Apply a little methylated spirit or *Dettol* on the skin and cover with a piece of leucoplast.

For serious cuts call the Doctor at once. In the mean while try to stop bleeding by applying pressure above the cut. The pressure should not be continued for more than five minutes.

Note. Minor bleeding can be stopped easily be applying concentrated ferric chloride solution or alum.

Eye Accidents

Acid in Eye. At once wash the eye with water a number of times. Then wash it with 1% sodium carbonate solution by means of an eye-glass.

Alkali in Eye. At once wash with water and then with 1% boric acid solution by means of an eye-glass.

Foreign Particles in Eye. Do not rub the eye. Wash it by sprinkly water into the eye. Open the eye and remove the particle by means of a clean handkerchief or cotton wool. Again wash freely with water.

Poisons : It a solid or liquid goes to the mouth, but is not *swallowed,* spit it at once and repeatedly rinse with water. If the mouth is scalded, apply olive oil or *ghee.*

Acids. Dilute by drinking much water or preferably milk of magnesia.

Caustic Alkalies. Dilute by drinking water and then drink a glass of lemon or orange juice.

Arsenic of Mercury Compounds. Immediately give an emetic *e.g.* One table spoon full of salt or zinc sulphate is a tumbler of warm water.

Inhalation of Gases : Pungent gases like chlorine, sulphur dioxide, bromine vapours etc. When inhaled in large quantities often choke the throat and cause suffocation. In such a case remove the victim to the open air and loosen the clothing at the neck. The patient should inhale dilute vapours of ammonia or gargle with sodium bicarbonate solution.

Fire

Burning Clothing. It clothes have caught fire then lay the victim on the floor and wrap a fire-proof blanket tightly around him. The fire in the burning clothes will thus be extinguished. *Never throw water on the person* as it will cause serious boils on his body.

Burning Reagents. In case of fire on the working table at once turn out the gas taps and remove all things which are likely to ignite. Following methods be used to extinguish the fire:

(i) If any liquid in a beaker or flask has caught fire, cover the mouth of the vessel with a clean clamp cloth or duster.

(ii) Most of the fire on the working table can be extinguished by throwing sand on them.

(iii) If any wooden structure has caught fire it is put up by throwing water on it.

(iv) Never throw water on burning oil or spirit. Since it will only spread the fire. Throwing of a mixture of sand and sodium bicarbonate on the fire is most effective.

A first aid box should be provided in every laboratory. It should contain the following materials.

Bandages (3-4 rolls of different sizes), gauze, lint, cotton wool, leucoplast.

A pair of forceps, a pair of scissors, safety pins. Glass dropper, two eye-glasses.

Vaseline, boric acid powder, sodium bicarbonate powder, a tube of Burnol.

Sarson oil, olive oil, glycerine.

Picric acid solution, Tannic acid solution 1% acetic acid, 1% boric acid, 1% sodium bicarbonate, saturated solution of sodium carbonate.

Methylated spirit, rectified spirit. Dettol.

Chemicals Stored

In storage of chemicals the following methods are normally adopted:

(i) Grouping the chemicals in a systematic way.

(ii) Arranging the elements in alphabetic order.

(iii) Arranging the elements and their components in which they occur in periodic table.

(iv) Grouping all elements and their similar components together.

(v) Numbering each bottle and jar and keeping and index book.

(vi) Keeping reagent bottles in definite places on the bench and the cupboard.

(vii) Storing similar types of solutions at one place.

(viii) Using coloured bottles or bottles with spots of coloured paint.

(ix) Using same type of bottles for a particular type of reagent.

(x) Always store large bottles on floor and not on shelves.

(xi) While storing Winchester bottles of concentrated acids, they be kept in brackets with sand.

(xii) Bottles containing inflamable liquids be stored in a cool place outside the laboratory.

(xiii) White phosphorus be stored under water and sodium be stored under kerosene oil.

(xiv) Hydrogen peroxide be stored in an air light tin.

Assistant's Role

Each laboratory be provided with a laboratory to perform the following duties:

(i) Keeping benches and laboratory clean.

(ii) Oiling benches with linseed oil.

(iii) To draft orders for chemicals and apparatus.

(iv) To receive the supplies of chemicals and apparatus after proper checking.

(v) To prepare solutions for volumetric analysis.

(vi) To prepare the solution for reagents shelf and to keep the reagent bottles full.

(viii) To set up apparatus for demonstration and experiments.

(viii) To maintain apparatus (burette, pipette etc.) in proper working conditions.

(ix) Keeping reagent bottles and chemicals at proper places.

(x) Periodic cleaning of iron stands, balances etc.

(xi) Keeping the first aid box replenished.

(xii) For repair of apparatus and glassware.

(xiii) Periodic checking of chemistry books in library and reference books in the laboratory.

Books for Consultation

Books containing following type of details be kept in the chemistry laboratory as reference books:

(i) Books containing physical constants of common substances.

(ii) Books which give the details of preparation of solutions for reagent bottles as also for volumetric analysis.

(iii) Recipes for cleaning glass ware and metals, removing stains, freeing glass stoppers, fire proofing etc.

The Regulations

Maintaining discipline in laboratory is more difficult as compared to maintaining discipline in the class room. This is so because, pupils doing the same work wish to talk and discuss with others. Modest talking is inevitable in the laboratory. Yet talking and walking in the laboratory may cause accidents. Following rules will help to avoid any such accidents:

1. Admission to the laboratory in the absence of teacher should be avoided.
2. Teacher should not be late unduly.
3. Students should silently go to their places after entering the laboratory.
4. Before beginning his lesson teacher should wait for silence.
5. The teacher should adress the whole class.
6. Teacher should see that a complete silence is observed during his talk.
7. Teacher should change his pitch at times to add interest to his talk.
8. Teacher should make adequate preparation to keep class busy.

9. The teacher should for see and remove all possible causes of trouble.

10. Adequate apparatus be made available.

11. The teacher should try not to produce rigid discipline.

Following rules should be observed by students working in a chemistry laboratory:

1. Nothing should be taken out of the laboratory without the permission of the teacher.

2. Apparatus and chemicals be used only for specific purposes.

3. Any mistake be brought to the notice of the teacher, immidiately.

4. Breakages be reported to the teacher.

5. Before returning the apparatus to its original place it should be thoroughly cleaned.

6. Wastage of water, petrol gas etc. be avoided.

7. Make economical use of chemical reagents and chemicals.

8. All doubts be cleared with the help of the teacher.

Following rules if observed will help to avoid accidents:

1. Never use concentrated acids unless specifically instructed.

2. Do not mix chemicals aimlessly.

3. Do not taste chemicals without permission.

4. Pour liquids only down the sink.

Well Equipped Laboratory

Laboratory work is an essential component of chemical education. However, schools in most developing countries are ill equipped for practical work, and even where laboratories have been built the problems of servicing the laboratory as well as equipment are more acute. The basic problem is of funding the purchase of consumables and equipment, together with the associated maintainance cost.

In one area at least, in the provision of equipment, solutions are possible. Hakansson has pointed out, if the selection of equipment for school science is based upon 'principles' rather than upon obtaining accurate results, very simple and quite in expensive equipment can be used.

Teacher should be very careful while planning the purchase of equipment for the laboratory. He should carefully weigh each item to be purchased with its educational worth. He may classify his requirements as under:

(i) Apparatus required for laboratory work.

(ii) Apparatus required for demonstration purposes.

(iii) Apparatus required for general use.

While determining the quantity of apparatus to be purchased, he should keep the following points in mind.

(i) Financial resources at his disposal.

(ii) Demonstration and laboratory work that has to be done during the year.

(iii) Scheme of work including the method of teaching to be used.

(iv) Storage facilities available.

Before making actual purchases a list be prepared of experiments to be informed by students and principles to be demonstrated by the teacher. For demonstration only one set will do but for experiments to be carried out by the students the number of sets required will be equal to the member of students working at a time. Some additional sets be purchased to cover up for the breakages etc. only such articles which are really required should be purchased.

If only limited funds are available then the purchases of beakers, flaskes, funnels, files etc. should be accorded first priority. These are the articles which are required by students while doing experiments. When ample store of such articles has been made then only the apparatus needed for demonstration be purchased. Another important point which must be considered is that there is enough accommodation for the proper storage of articles likely to be purchased. While ordering for purchases scheme of teaching, method of teaching and knowledge and ability of teacher must also be given due consideration.

Keeping in view the points discussed above the teacher should prepare a list of articles to be purchased. While selecting apparatus teacher must not be tempted by attractive descriptions given in catalogue. While preparing an indent the teacher should give full specifications of the article required. In the absence of such specifications it is just possible that you make purchase of items which you never intended to purchase. For selecting a good firm the list of apparatus with complete specification be sent to some competing and reputed firms and they be asked to quote their lowest rates. A specimen for inviting quotation in shown below.

After receiving quotations a reliable firm quoting the lowest rate be asked to make the supplies.

The UNESCO designed apparatus for tropical schools is very satisfactory. The government of India is considering an arrangement for the manufacture of such instruments and making supplies of these to schools in lieu of cash grants.

S. No.	*Articles*	*Quality*	*Size*	*Quantity*
1.	R.B. Flasks	Pyrex glass	250ml	10 doz.
2.	Titration flasks	Pyrex glass	100ml	10 doz.
3.	Troughs	Pneumatic glass	12"dia	2 doz.
4.	Nitric acid	Commercial		51
5.	Sulphuric acid	B.D.H.		11

Chemistry teacher should carefully check the items received and then arrange them properly after making entries in the stock register.

Availing the Equipment

After preparing the list of items to be purchased the chemistry teacher should make purchase from the firms approved by the controller of stores or from some other approved source. However if no such source is available then he should send a list of his requirements to reputed firms for quotations. After receiving quotations, the teacher can select the dealers and firms for placing orders—the criterion being lower price and better quality.

It is desirable to patronise local and neighbouring firms. This provides the teacher a chance to select personally the items required and get them packed in his presence. Many a times it is better to go to a firm of repute and make the selection and purchase, and have the apparatus packed in one's presence. It is always wise to get insured against breakage and loss of the fragile apparatus ordered from out station firms.

On receiving the apparatus it should be carefully unpacked and after proper checking all the items of the purchase be catalogued and recorded in stock register then the goods received be properly stocked.

The Management

The apparatus received be arranged in almirahs provided with glass fronts and preferably be fitted with mortise locks to avoid dust getting in. Apparatus should be arranged in such a way that each and every item could be easily located. Items which are frequently required be stored at such places that they are readily available. The apparatus should be arranged one deep of shelf; several rows of same articles may be placed on one shelf. The apparatus may be arranged either subject wise or alphabetically. If we arrange the apparatus subject-wise we find that some articles fall under more than one heading and if we arrange them alphabetically we find that glass and metal articles are coming together in which there is more likely hood of breakage. Thus it is always better to reserve a few almirahs for apparatus required for individual practical work in which the apparatus be arranged alphabetically. The chemicals can also be stored alphabetically.

A list be pasted on the almirahs showing the names of articles stored in it.

The apparatus needed only for demonstration purposes may be stored in separate almirahs subject-wise.

Chemicals can be stored even on open shelves. For this purpose two open shelves can be provided on either side of the recess for balances. However dangerous and costly chemicals like phosphorus or sodium or salts of mercury, bismith or cadmium be stored separately in an almirah. The containers or bottles containing chemicals should be neatly labelled.

To effect economy in space some items of common use such as stands, holders, clamps etc. many be stored outside almirahs. Two possible arrangements for storage of iron stands.

The Maintenance

Care of equipment and apparatus is one of the important functions of the chemistry teacher. For this the apparatus kept in almirahs must be checked at regular intervals. During this checking operation the apparatus should not only be inspected but it should also be dusted, cleaned and polished if necessary. If proper care is taken the life of the apparatus will increase. For proper upkeep and maintenance the following points be kept in mind.

(i) After use the apparatus should be properly cleaned before it is returned to its proper place. Never return dirty apparatus to its proper place. This is specially applicable in case of glass apparatus used in chemistry laboratory. For proper cleaning of glass articles we can make use of soap, pumice stone, hot alkali solution, acidified potassium dichromate solution etc.

(ii) For cleaning items made of brass we can use *Brasso*. *Brasso* be applied to the article with finger covered with a piece of muslin, allowed to dry and then rubbed off with a clean duster.

(iii) Iron articles are generally polished. For polishing such articles use Black Japan thinned with a little turpentine or kerosene oil. Aluminium paint can also be used. If the article to be painted has any rust it should be removed, by rubbing with an emery paper, before painting the article. Use of kerosene oil can also be made for removal of rust. Take care to apply vaseline on screws and hinges of iron articles during rainy season.

(iv) Wooden articles be left in the sun after being polished with spirit polish. Spirit polish can be made by dissolving shellac in methylated spirit. One or two coatings of it are then applied on the article.

(v) The top of each laboratory table is unpolished but it should be waxed (specially in case of chemistry laboratory) to avoid the action of acids. For waxing either paraffin wax or candles can be used. Wax is coated over the table with the help of painters brush and is then spread over and smoothened using the hot iron of washerman. It is then allowed to dry and any excess of wax is scarped off with a blunt knife. It is then polished with a coarse duster.

(vi) Special attention be paid to keep sinks clean. For cleaning sinks use vein powder or some other cleaning powder. Use special chemicals for removal of stains if they persist.

(vii) In chemistry laboratory special attention be paid to the fact that stoppers of bottles are not lost or get changed. For this they should be tagged to the bottle either using a copper wire or a rubber band.

(viii) The apparatus which is frequently used by students may go out of its proper adjustment and a good chemistry teacher must find time for its proper adjustment and must also be able to carry out minor repairs. For this the chemistry laboratory must be equipped with a tool kit containing usual hammer, wrenches, pliers, screw drivers, forceps etc.

(ix) Glass panes of almirahs should also be cleaned occasionally. For cleaning glass panes use monkey brand soap. Rub a wet sponge over the soap and then over the pane and clean off with a duster. Pumice stone dipped in water is in methylated spirit and rubbed over the panes will remove all dirt.

(x) For cleaning of glass apparatus in general and burette and pipette in particular, use a solution of potassium dichromate acidified with dilute sulphuric acid.

Detailed Accounts

Maintaince of a proper record of the apparatus, material etc. in the laboratory is one of the important duties of the science master. For this after receipt of articles they should be thoroughly checked and then they be entered in the stock register. A specimen page from a stock register is shown on next page.

Separate stock register be maintained for consumable and non-consumable items, permanent articles, glass articles etc. Following stock registers are generally maintained in schools:

(i) Stock register for non-breakable articles.

(ii) Stock register for breakable articles.

(iii) Stock register for consumable articles.

(iv) Stock register for permanent articles.

In addition to various stock registers following registers should also be maintained.

Order Register : This Register is meant for orders sent for the purchase of new apparatus. Entries in this register should indicate the serial number and date of the order, name of the firm to whom the order has been placed, details of articles ordered, articles received, cost of articles received. For convenience a copy of the order be posted on the left hand page of this register and a copy of supply under be pasted on the right hand page.

Requirement Register : This register if maintained makes the task of placing orders easier. Teacher will enter in this register the items whose absence is felt by the teacher at the time of demonstration of an experiment or during the practical class. If such entries are not made them there is every likelihood that some of the items needed by the teacher may be left out while placing the order for purchase *of* material and equipment. The requirement register should invariably be consulted by the teacher whenever be places orders for the purchase of materials or other requirements of the laboratory.

Specimen Page from Page............

Stock Register

Stock Register of....................................Department

Name of the article...............................

Month & Date	*Particulars (Name of the Firm, Bill No. and Date)*	*Receipt written of*			*Consumed/*		*Balance*		*Initials of Tr. incharge*
		Qty.	*Rate*	*Amt.*	*Qty.*	*Amt.*	*Qty.*	*Amt.*	

Stock Register for Science Club : Organising science clubs is quite useful in creating a scientific atmosphere and each secondary school is expected to have a science club. For organising various activities of science club the teacher needs different types of apparates, equipment and materials. It is desirable to maintain a separate stock register for the science club. In this register all the apparatus meant for science club be entered. Entries of models, charts and collections made by students should also be made in this register.

Maintainance of Stock Registers. Following points should be given done consideration while maintaining any stock register:

(i) The outer cover of the stock register should indicate the name of the register, name of the school, date of opening and closing of the register, etc.

(ii) A certificate be given on the first page of the register indicating the total number of pages in the register. All the pages be numbered serially and the above certificate be countersigned by the head of the institution.

(iii) Either seperate stock registers be maintained or the same stock register be divided into a number of portions under various heads such as Mechanics, Heat, Light, Sound, Electricity, Magnetism etc.

(iv) An effort be made to make entries alphabetically.

(v) An Index be given at the beginning of the stock register.

(vi) Each receipt entry should be entered with date of receipt and the items consumed or broken be shown in the columns meant for this purpose. All these entries be initialled by chemistry teacher and countersigned by head of institution.

It is expected that science teacher is capable of devising and making apparatus for some simple experiments, modify apparatus and carry out simple repairs. The apparatus devised and made in school work shop or laboratory by the teacher or student is known as home-made apparatus. A science teacher with a little thought and ingenuity can make a number of valuable and serviceable models making use of cheap materials such as Jam-Jars, bits of wire, corks, motor parts etc. Some of the advantages of using home-made apparatus are as under:

(i) Such an apparatus is economical.

(ii) Use of such an apparatus makes more obvious the application of science to life and things around us.

(iii) It provides an encouragement to the student to make such an apparatus and adopt it as a hobby.

(iv) It helps to correlate science with manual training.

(v) It creates extra interest in the subject.

(vi) It provides training in manual skill, resourcefulness and ingenuity. These qualities are quite useful for life.

A *word of caution* for teacher in using house-made apparatus is that he must not sacrifice efficiency just for his over enthusiasm for using house-made apparatus.

Such equipment can be made by individual teachers for their own use in schools or made available from a production centre. This type of equipment can serve the needs of the teacher, the student and the curriculum more effectively. As already pointed out such an equipment can be produced by individual teachers or can be procured from production centres.

Different Equipments

The UNESCO source book for Science Teaching contains a number of suggestions for simple teacher-made equipment in addition to a wide variety of experiments. One chapter has been specially devoted, in UNESCO Handbook for Science Teachers, to facilities, equipment and materials. The Guidebook to Constructing. Inexpensive Science Teaching Equipment, which have been produced at the university of Maryland (United States).

A phamphlet has been produced by the Junior Engineers/ Technicians/Scientists (JETS) based in the school of Engineering, University of Zambia. It is intended to help schools produce equipments such as wooden racks and stands for pipettes, burettes and test-tubes, and metal clamps, clamp holders and retort stands.

In India similar work has been undertaken by the National Council for Educational Research and Training (NCERT). Details for a mobile laboratory unit has been published in India.

The *Manual de quimica experimental,* produced in Bolivia contains a number of experiments which illustrate most of junior secondary level chemistry course. *e.g.* preparation and properties of common gases; acids, bases and salts; laws of chemical composition. In this manual instructions are written for teachers with little or no workshop experience, on how to make simple balances, various supports, an alcohol burner and some items of electrochemical equipment. It also provides a list of chemical that can be procured locally from market or pharmacy.

The production of equipments by teachers in their own schools and its advantages were taken up in the previous section. However many a teachers find it burdensome because of the fact that they are faced with day-to-day difficulties of teaching. Really speaking it is too much to expect teachers to be the sole providers of equipment. From their efforts we can develop local production units and the teachers can then be expected to maintain the equipment supplied to them. Warren and Lowe's. *The Production of School Science Equipment* provides an insight into developments in various countries. A summary of experience in Bangla Desh, Figi, Pakistan, India, Hongkong, Japan, Indonesia, Philippines, Singapore, Vietnam and Republic of Korea has also been published.

There are some large-scale projects, in developing countries, for production of locally based equipments. The concept of centres is not new, a prototype being set up in chile in 1964. Other production centres are NCERT (New Delhi), IPTST (Bangkok), the Science Education Production Unit (SEPU) in Kenya and the National Educational Equipment Centre (NEEC) in Pakistan.

NCERT (New Delhi) makes batches of 1500 lots for primary and middle schools and is under contract to UNICEF for 50,000 kits.

SEPU produces teaching aids *(e.g.* Slides and Photographs) and chemistry, biology and physics kits for secondary schools which are designed to meet all the practical requirements associated with East African Certificate of Education. Kits are accompanied by manuals for teachers and students. The emphasis is on pupil participation and small-scale experiments thus the kits are not suitable for demonstration work.

Some of the most essential points that must be kept in mind while establishing production centres for low-cost equipment are summarised below:

(i) The centre must have *expertise in design, in management* and *distribution.*

(ii) To overcome the *shortage of technicians,* the training of management staff and training of technicians is of vital importance. The committee on the Teaching of Science of International Council of Scientific Unions (ICSU-CTS) in conjunction with UNESCO, is endeavouring to discover the extent of the shortages and find ways of alleviating them.

(iii) *Realistic Budgeting.* The production centre must work to realistic budgeting. To lower the cost of production the production centre must be cost-concious. Making as large a range of apparatus as possible from a given item of equipment will help to lower production costs.

(iv) *Effective Marketing and Distribution.* For effective marketing and distribution it is essential to make an infrastructure between the production centre and the educational establishments. In small countries production centres may be set up to serve both the schools and the institution of higher studies.

(v) *Cooperation with Teachers and Curriculum Designers.* The production centres should design the equipment, to be produced, in conjunction with teachers and curriculum designers and only such equipment as needed in view of the requirements of prevailing text-books be only produced.

(vi) *Quality Control.* Before supplying the equipment to schools it must be checked for the quality including reliability. Only good quality equipment be marketed.

(vii) *Facility for Repairs and Maintainance.* The production centre must have an efficient system for repairs and maintainance.

The Material

It is possible to reduce the cost of teaching a laboratory-based chemistry curriculum by using small scale techniques. It is also important to consider how much and what chemicals are to be used. Small-scale techniques are generally more safe and they also help to improve the manipulative skills of the students. Texts indicating how small-scale work can be used through out a school course have been published in many a countries.

To further reduce the cost of materials it is desirable that locally available chemicals are put to maximum use *e.g.* geochemical minerals, disused dry cells, scrap metal, vegetable oils, orange peel, root extracts, soap and baking powder.

In Thailand, IPTST has produced a detailed list of chemicals readily available in local markets. The production of similar lists by institutions in other countries would be of much use.

Role of Computers

We find that in new programmes in chemistry teaching the computers are used increasingly. Many articles that have appeared in literature also point to the increased use of computers in teaching of chemistry. The applications of micro-computers in school chemistry can be classified as under:

Direct Teaching. In this type are included the use of computers for simulations, instructional games, revision questions and exercises.

Data Handling. This includes word processing, data base management and data collation and display in the laboratory.

Computer Assisted Learning. From the survey of literature we can easily find that most commonly the micro-computers are used in teaching of chemistry are used for handling of experimental data, *e.g.* interfacing with a gas chromatograph, monitoring and controlling clock reactions and the calculations of numerical constants.

Though micro-computers are quite expensive yet their education potential is considerable and they offer chemistry teachers an opportunity to experiment with imaginative and innovative ways of teaching chemistry.

Various Devices

Charts, diagrams, pictures etc. if displayed in the laboratory provides right scientific atmosphere to the place various details about these have been discussed in chapter on *Teaching Aids*. Here the topic is discussed just as a reference.

Charts : An all out effort be made to avoid display of printed charts available in the market because these charts are quite costly and are not fully representative. Such charts are also sacrifice simplicity and directness to details.

Following type of charts be preferred for display in chemistry room:

(i) Charts showing diagrammatic sketches of different pieces of apparatus generally used by students in their practical work. e.g. beaker, flask, gas-jar, retort, spirit lamp etc.

(ii) Charts depicting diagrammatic sketches of different important experiments from various branches of chemistry e.g. chart showing the preparation of oxygen, hydrogen, carbon di-oxide etc. Such a chart should be fully labelled and should be drawn in lead pencil.

(iii) Some charts for use in demonstration lessons.

(iv) A progress chart depicting the progress of each student be prominently displayed. Such a chart should show the complete record of work of the student.

(v) Some important do's and don'ts be also displayed on a chart placed at some prominent place in the laboratory.

(vi) A chart of common accidents and first aid be also depicted in the laboratory.

In addition to various types of charts given above, the following types of *pictures* and *illustrations* are quite useful if depicted in the laboratory:

(i) Portraits of great Indian and world chemists.

(ii) Pictures of scientific interest *e.g.* pictures of Nangal Fertilizer Project.

(iii) Pictures showing progress of chemistry, *e.g.* pictures of atomic power stations.

(iv) Maps indicating sources of ores of metals and chemical products.

(v) Weather charts, maps and graphs prepared by students after observing and collecting data from weather reports.

(vi) Various types of demonstration models preferably prepared by students.

Each laboratory is expected to have at least three boards to be used as bulletin boards. These are to be used as under:

(i) One of the boards is reserved for display of newspaper cuttings, sciences news and pictorial illustrations of scientific interest.

(ii) One of the boards is reserved for putting up notices about science club activities.

(iii) One of the boards may be used for indicating the assignments.

For teaching of chemistry availability of good apparatus and well equipped laboratories is a must. However it should lead us to a wrong conception that teaching of science cannot be carried out in the absence of expensive apparatus. One of the reports by NCERT observes that from among various factors that stand in the way of science education in our country one is lack of adequate resources for laboratory building, purchase of good and adequate apparatus and equipment. This lack of funds and resources makes improvisation of apparatus almost a necessity in India.

Need for Improvisation : India is a poor country and so we have only limited financial resources. For imparting effective and efficient science education, due to this financial constraint we require the production of improvised and inexpensive learning aids. A teacher with some ingenuity and manual skill can make a number of valuable and serviceable articles from discarded things all around him. For this purpose every science room should be equipped with a work bench and a kit of tools that may be used by students and teacher in making and improvising equipment for chemistry teaching.

Definition of Improvisation : Some of the definitions of improvisations are given below:

It refers to a make shift arrangement for accomplishing the intended learning task.

It refers to contrived situation that is created from reading available material for sake of convenience.

It refers to a stimulating situation for demonstrating and imparting learning is respect of controls and operations making use of low cost materials.

It refers to those learning aids which are prepared from simple and readily available cheap material by students and teacher.

Significance of Improvisation : Improvisation is quite significant and has many values as the process of improvisation needs resourcefulness and ingenuity on the part of the chemistry teacher. It is based on the concept of solving some problem by a make shift or alternate arrangement given below are some significant values attached with the process of improvisation:

(i) It splashes the cost of apparatus and is quite helpful in making the school self-reliant.

(ii) It has instructional value as well. When we are carrying out any improvisation we do get a proper feeling for the scientific process and designing. Thus we learn by doing.

(iii) It help develop the dignity of labour and also satisfies the urge of creative production.

(iv) It helps to develop the habit of cooperation and coordination.

(v) It provides training in thinking skills through the process of looking for low-cost substitutes or alternatives.

Process of Improvisation : It refers to a systematic way of constructing a piece of apparatus or designing an experiment. It involves the following steps:

(i) Making a careful study of the conventional apparatus or experiment.

(ii) Thinking of some low cost substitute that may be available in the market.

(iii) Designing the improvised apparatus or experiment.

(iv) Putting the improvised apparatus or experiment to test.

(v) Making further improvements in the improvised apparatus keeping the test results in mind.

(vi) Making use of the improvised apparatus in the laboratory for demonstration or practical work.

Examples of Improvised Apparatus : Some examples of improvised apparatus are given below.

Simple Tripod Stand : To make a simple tripod stand we have only to cut away A or n shaped piece from the sides of a discarded tin can. We can remove the lid and bottom of the tin can completely or we can simply make holes in the bottom. If holes are made it also serves the purpose or wire gauze.

Beehive Shelt : An improvised beehive shelf can be obtained from empty tin can. The tin can to be used for the purpose should be rust free and its inner and outer surfaces are either galvanised or varnished. To make a beehive shelf drill a hole of 1/2″ diameter in the centre of the bottom of tin can and cut a V-shaped notch on one side of it. This can now be used as a beehive shelf.

In this boiling water from a kettle is allowed to condense in a jam jar which is immersed in a pan containing ice cold water. A simple glass tube fitted with a rubber tubing can be fitted to the mouth of the kettle and another glass tube is fitted to serve as outlet for condensed steam.

There are many more such items which can be easily obtained. Some such items are:

(i) Spring balance.

(ii) Spirit lamp.

(iii) Water voltameter.

(iv) Fire extinguisher.

Advantages of Improvised Apparatus : Some of the advantages of improvised apparatus are:

(i) These are quite cheap and economical.

(ii) They have great educational value. While devising such apparatus students gains more familiarity with the underlying principles of the apparatus.

(iii) It helps to develop the creative and constructive instructs of the child.

(iv) It inspires young students to explore and invent new things.

(v) It develops the lower of initiative and resourcefulness in the student.

(vi) It helps to develop power of scientific thinking.

(vii) It helps to inculcate the habit of diligency in the students.

(viii) It galvanises dignity of labour.

(ix) It solves problem of leisure time.

Special Items

A chemistry kit is a container or box in which are stored some special apparatus/material and so these are readily available for experimentation and demonstration work.

The central science work shop, a wing of NCERT, New Delhi, has prepared kits for teaching of chemistry to various classes alongwith manuals and detailed guidance for the use of these kits by the teacher and the students.

The material stored in these kits is easily available in market and some of the items of these kits can be improvised or made in work shop by carpenters or black smiths. These kits are of three types:

(i) Kits which are helpful for demonstration of certain specific facts and principles.

(ii) Kits which help the students in performing experiments and draw inferences so as to study facts and principles.

(iii) Kits which are useful for both teachers and students.

The *demonstration kit* devised for chemistry by NCERT can serve the purpose of demonstration kit for classes VII and VIII. It contains chemicals, glass apparatus etc. required for demonstration of various experiments. In it there is a total of 63 items of apparatus, 67 chemicals and it contains 12 types of containers. Apparatus stored in the kit is mostly improvised apparatus and the containers are bottles and ink phials. All the items of the kit are arranged in a suitable manner and are packed in a box made of wood. Such a box can be easily transported. This kit contains no liquid chemicals and these have to be arranged locally whenever the need arises.

Chemistry pupil's kit is meant for use of students in their practicals and it can serve as a 'mini laboratory' for village schools. The front covers of the kit can be hinged to serve as a table. It contains 37 different items, 40 chemicals and 10 types of containers for chemicals. It is also provided with two test tube stands, one on each side of the box. These can also be used as handles for lifting the box.

The Advantages : Some of the important advantages of the science kits are summarised below:

Low Cost. These kits are very economical and a primary science kit costs only around rupees one hundred. All the kits for a middle school are not likely to cost more than Rs. 3000/-.

Use of Indigenous Resources. Majority of items stored in chemistry kits are indigenous and no import or foreign collaboration is required.

Easy Replacement of Items. Various items stored in chemistry kit can be easily replaced in case of loss or breakage etc.

Easily Portable. The kits are easily portable and can be easily used for demonstration in classes being met in open or under trees.

Economy of Time. Being planned in a very systematic way, minimum time is needed for setting up an experiment by using such kits.

Consolidated Material. All the essential items of apparatus, equipments, chemicals etc. are arranged in such a way that any item can be put to multipurpose use. In this way material is consolidated in a chemistry kit. Moreover all the material needed for an experiment is consolidated in the kit.

Economy in consumption. In a science kit the items stored are in mini size and so the consumable items such as chemicals etc. are consumed in small amounts.

Teacher's Innovation. The science kits can serve to motivate and inspire an innovative chemistry teacher to work out new ideas and he can encourage his students to improvise new apparatus and experiments.

Students Involvement. Since the various items stored in science kits are quite simple so they can be easily handled by the students. It encourages the students and they become active participants in the teaching-learning process.

Value of Textbooks

In the present educational set up the role of text-book is of prime importance. However we find that little attention is paid to this important aspect of education. Most of the text-books in chemistry are not of good standard. They follow the prescribed syllabus too rigidly and no attention is paid to develop the topics according to the need and interest of students. A good text book is one which is a source of knowledge arranged systematically ands it enables the reader to acquire the needed information quickly. It inspires the student to invent, to discover and to inculcate scientific methods. However teacher should not depend solely even on the best of the text books because even such a text book omits many details which teacher wants to tell to his students.

The use of a text-book is made by the students for completing the preparatory part of an assignment. They also use their text-book for doing revision of course. Some students also consult and use their text books to study at home, the demonstration lesson given to them by their teacher in school. In this way text-books are used to supplement the class work. Text books also provide a help to students in correct understanding of basic concepts and principles of science.

Generally a numbei of books are prescribed by board or university to be used as text books. NCERT prepared text books are available upto class XII. While recommending a text-book to his students the teacher should consider the following points to assess the worth of the book:

(i) Correctness of matter.

(ii) Purity of languages.

(iii) Simplicity of diagrams.

(iv) Quality of printing and binding.

Correctness of Matter : In this correction the standing of the author and the reputation of publishers should be considered. The books written by well known author having a long teaching experience of teaching the subject and possessing requisite qualifications be recommended. It would be much appreciated if certain minimum qualifications and experience for authors is laid down by authorities.

Purity of Language : A text book that presents the subject matter in a simple, clear and lucid language should be preferred. For text book in a regional language, the scientific terminology should also be given in English with in brackets. In such books only standard terminology evolved by the Central Ministry of Education and State Governments should be used.

Simplicity of Diagrams : Only simple and well labelled diagrams be given in text books. Such diagrams are self explanatory and help the student in properly-understanding the subject matter.

Quality of Printing and Binding : It is desirable that a text book makes use of a good quality paper and the quality of printing, and type of letters in fine. It should be so bound that its binding is appealing to the student.

In addition to the above a good text book is expected to select and arrange the subject-matter in a psychological sequence. The book should follow the aims of teaching chemistry and should serve as a guide for demonstration lesson as also for individual experiments. Each chapter should start with a brief introduction and a summary of the subject matter be given at the end of the chapter. Some assignments should also be given at the end of each chapter and the assignments should cover such areas as applications to life situations, numerical questions, suggestions for experimental work and projects, objective type tests etc. Heading and sub-headings be given in bold type. A table of contents be provided at the beginning and a subject-index be provided at the end. Glossary of some important scientific terms be given at the end of the book.

In most of the school still the traditional school chemistry text books are in use. Such books are used for reading, revision, to follow instructions for some laboratory exercise etc. However traditional chemistry text books have been abandoned in those countries in which 'process' has become a significant feature of school chemistry courses. In such countries generally well illustrated work-books that contain exercises for fostering the development of such 'process skills' as observing, hypothetising and inferring. These are accompanied by data books and books of questions for pupils and teacher's guides.

Text books also communicate to pupils various assumptions about the nature and purpose of scientific activity.

12

Study of Gases and Acids

Since pupils have to learn chemistry and thus content of chemistry is to be given to the students. Thus for chemistry learning the content should be as good as the method of teaching. It is with this view in mind that some content portion is assigned to the syllabus for teaching of chemistry. In the pages to follow we will take up certain concepts in chemistry.

Oxygen

It is present in the air (atmosphere) in the free (native) form and it is about 20% of air by volume. Lavasior detected the presence of oxygen in atmosphere. Sheele obtained the gas in laboratory and studied its properties.

Preparation : In the laboratory oxygen gas is prepared by heating, a mixture of potassium chlorate (4 parts) and manganese dioxide (1 part), in a hard glass test tube fitted with a delivery tube.

The. other end of the delivery tube is placed under beehive shelf kept immersed in a through of water. Over the beehive shelf is placed an inverted gas cylinder filled with water (Fig). On heating test-tube gently the gas bubbles can be seen rising in the gas and the gas is collected by downward displacement of water. The chemical reaction taking place can be represented as under

$$2\ KClO_3 \xrightarrow[\text{(catalyst)}]{MnO_2} 2KC1 + 2\ O2\uparrow$$

Precautions

(i) Always use pure manganese dioxide.

(ii) Before removing the flame remove the delivery tube from the beehive shelf or water as the water may rush into the test-tube resulting into its breakage.

(iii) The test-tube should be clamped in the stand in a slanting position to avoid breakage of the tube by the condensed vapour.

(iv) Test-tube should be heated gently and slowly.

(v) A glass lid should be placed on the mouth of the jar after filling it with the gas. It may be made air tight by applying a little glycerine or vaseline on its surface.

Properties of Oxygen Gas (Physical Properties)

(i) It is a colourless, tasteless and odourless gas.

(ii) It is slightly soluble in water. The dissolved gas is used by the animals living in water for respiration.

(iii) It is slightly heavier than air.

(iv) It can be liquified by lowering the temperature and increasing the pressure.

Chemical Properties

1. It is neutral to litmus.

2. It is not combustible but it is a supporter of combustion.

3. It reacts with hydrogen under the influence of an electric spark and produces water

$$2H_2 + O_2 \xrightarrow[\text{Sparks}]{\text{Electric}} 2H_20$$

4. It reacts with metals its form their oxides.

$$2\,Mg + O_2 \rightarrow 2\,MgO$$

$$4\,Na + O_2 \rightarrow 2\,Na_2O$$

$$4\,Fe + 3O_2 \rightarrow 2Fe_2O_3.$$

5. It oxidises ammonia to nitric oxide.

$$4NH_3 + 5O_2 \xrightarrow{\text{Pt, 800°C}} 4NO + 6H_2O$$

6. It can be converted to ozone (ozonised oxygen) by passing electric sparks at ordinary temperature and pressure.

$$3O_2 \xrightarrow[\text{Discharge}]{\text{Electric}} 2O_3$$

7. A mixture of acetylene and oxygen when burnt produces a very hot flame which is used in welding metals.

$$2\,C_2H_2 + 5O_2 \rightarrow 4CO_2 + 2H_2O$$

Uses.

1. It is used as an oxidising agent.
2. It is used in welding.
3. It is used in artificial respiration.
4. It is used in preparation of ozone.

Hydrogen

It is the lightest element and also the lightest gas. It was discovered by Heavy Cavandish in 1663. The name hydrogen was given by Lavasior in 1783.

Preparation : It can be prepared from acids, alkalies, water etc.

In the laboratory hydrogen gas is prepared by the action of zinc with dilute HC1 or dil H2SO_4. The chemical reactions taking place can be represented as

$$Zn + H_2SO_4 \rightarrow ZnSO_4 + H_2\uparrow$$

$$Zn + 2HC1 \rightarrow ZnCl_2 + H_2\uparrow$$

A Woulfs bottle is taken and some pieces of granulated zinc are placed in it. Then a thistle funnel is fitted in one mouth and in the other mouth a delivery tube is fitted. Some water is added to cover the zinc pieces. Then conc. H_2SO_4 or HCl is through the funnel. The hydrogen gas coming out of the Woulf's bottle through delivery tube is collected by downward displacement of water.

Precautions

1. The appartaus should be made air tight.

2. The lower end of thistle funnel must be under water in the Woulfs bottle.

3. No flame be allowed near the apparatus.

4. Pour the conc. acid slowly in the Woulfs bottle.

5. Gas is always stored in inverted gas cylinders.

Properties (Physical Properties)

1. It is a colourless, odourless and tasteless gas.

2. It is lighter than air.

3. It is insoluble in water.

4. It can be liquified by decreasing the temperature and increasing the pressure.

Chemical Properties

1. It is a combustible gas.

2. It explodes in presence of air.

3. It is neutral to litmus.

4. On being burnt in oxygen, it forms water.

$$2H_2 + O_2 \rightarrow 2H_2O$$

5. It combines with halogens to yield the corresponding halides.

$$H_2 + C1_2 \rightarrow 2HC1$$

$$H_2 + Br_2 \rightarrow 2HBr$$

$$H_2 + I_2 \rightarrow 2HI$$

6. It is a strong reducing agent.

$$CuO + H_2 \rightarrow Cu + H_2O$$

$$Fe_3O_4 + 4H_2 \rightarrow 3Fe + 4H_2O$$

Uses

1. It is used as a reducing agent.
2. It is used in preparation of ammonia, methyl alcohol, hydrochloric acid etc.
3. It is used in preparation of Vanaspati Ghee.
4. It is used in welding (Oxy-hydrogen flame)

Carbon-Di-oxide

CO_2 is produced during the respiratory by all living beings including vegetable kingdom. During the day in the presence of sunlight plants absorb carbon-dioxide and give out oxygen gas. In this way carbon cycle is formed to keep its balance in nature.

Preparation of Carbon-Di-oxide : It can be prepared by the action of an acid on a carbonate or a bicarbonate.

In the laboratory carbon-dioxide is prepared by the action of marble or chalk ($CaCO_3$) with dilute HC1 or H_2SO_4.

$$CaCO_3 + 2HC1 \rightarrow CaCl_2 + H_2O + CO_2 \uparrow$$

A Woulf's bottle is taken and some pieces of marble, chalk or shells are put in it little of water is added to cover the marble. A thistle funnel and a delivery tube bent at-right angles is fitted.

Acid is added through the thistle funnel. The reaction occurs. The gas is collected by upward displacement of air. The gas is not collected over water because the gas is highly soluble in water. The gas can be tested with the help of a burning splinter. It extinguishes a burning splinter or a match stick.

Precautions

1. The apparatus should be air tight.
2. The lower end of the thistle funnel should remain dipped in the acid contained in the Woulfs bottle.
3. The marble pieces should be completely immersed in dilute hydrochloric acid.
4. Use dry cylinders for collecting the gas.

Physical Properties

1. It is a colourless gas.
2. It has a characteristic smell.
3. It is heavier than the air.
4. It is soluble in water.
5. It is acidic in nature.

Chemical Properties

1. It is neither combustible nor a supporter of combustion. Burning objects get extinguished in carbon-dioxide gas.

2. Some metals such as magnessium, sodium, potassium continue to burn in carbon dioxide gas while carbon is set free

$$2Mg + CO_2 \rightarrow 2MgO + C$$

$$4Na + CO_2 \rightarrow 2Na_2O + C$$

3. When carbon-dioxide dissolves in water, it forms an acidic solution which turns moist blue litmus paper red.

4. It reacts with alkalies to form carbonates

$$2NaOH + CO_2 \rightarrow Na_2CO_3 + H_2O$$

$$CaO + CO_2 \rightarrow CaCO_3$$

5. When CO_2 is passed through lime water *i.e.*, $Ca(OH)_2$, it turns it milky

$$Ca(OH)_2 + CO_2 \rightarrow CaCO_3 + H_2O$$

If we continue passing CO_2 gas in lime-water in excess, it again turns colourless

$$CaCO_3 + H_2O + CO_2 \rightarrow Ca(HCO_3)_2.$$

6. On being passed over red hot coal, it is reduced to carbon mono-oxide.

$$CO_2 + C \rightarrow 2CO\uparrow$$

7. Carbon-dioxide gas is absorbed by green plants in the presence of chlorophyl, sunlight and water to form glucose, starch, sugar or cellulose. This process is called photosynthesis

$$6CO_2 + 6H_2O \xrightarrow[\text{Sunlight}]{\text{Chlorophyl}} C_6H_{12}O_6 + CO_2\uparrow$$

Uses

1. Carbon-dioxide is used in preparing aerated water.

2. It is used in the manufacturing of solid carbon dioxide called dry ice.

3. It is used in the manufacturing of baking soda and washing soda.

4. It is used to neutralize the effect of lime in sugar industry.

Acids and Bases

Important concepts (theories) of acids and bases are proposed by:

(i) Arrhenius (1887)

(ii) Bronsted-Lowry (1923)

(iii) Lewis (1923)

Arrhenius Concept : Arrhenius (1887) defined acid as a substance that will dissociate to yield a hydrogen ion while base in one that will dissociate to yield a hydroxyl ion in aqueous solution.

Thus:

$$HCl\,(aq) \qquad H^+(aq) + Cl^-(aq)$$

Acid

$$NaOH\,(aq) \qquad Na^+(aq) + OH^-(aq)$$

Base

According to this concept HNO_3, HCl, H_2SO_4, CH_3COOH etc. are acids and NaOH, KOH, NH_4OH etc. are bases.

This definition is of limited application and is applicable in aqueous solution only. It does not cover those substances which fail to give H^+ or OH^- ions but behave as acids or bases.

Bronsted-Lowry Concept : According to this concept an acid is a substance that can donate a proton and a base is a substance that can accept a proton *e.g.*

HC1 (aq) + H_2O(1)	H_3O^+ (aq) + Cl^- (aq)
Acid Base	Acid Base
NH_4 + (aq) + H_2O (1)	H_3O^+ (aq) + NH_3 (aq)
Acid Base	Acid Base
H_2O (1) + NH_3 (aq)	NH_4^+ (aq) + OH^- (aq)
Acid Base	Acid Base
H_2O(1) + CO_3^{2-} (aq)	HCO_3^- (aq) + OH^- (aq)
Acid Base	Acid Base

It may be noted that an acid after losing a proton becomes base where as a base after accepting the electron becomes an acid.

A base formed by the loss of proton by an acid is called *conjugate base* of the acid. An acid formed by the gain of proton by a base is called *conjugate acid* of the base. Acid-base pairs such as H_2O/OH^-, NH_4^+/NH_3 etc. are called *conjugate acid-base pairs.*

Those substances which can act both as an acid and a base are called *amphoteric substances.*

It is important to note that (i) all Arrhenius acids are Bransted acids but all Arrhenius bases are not Bransted bases, and (ii) Bransted-lower concept is not limited to molecules to act as acids and bases but ionic species may also be considered as acids or bases.

This concept serves well in protonic solvents like water, ammonia, acetic acid etc. but fails in case of some obvious acid-base reactions *e.g.* it can not explain how acidic oxides such as an hydrous carbon dioxide, sulpher dioxide, sulphur trioxide etc. neutralize basic-oxides like calcium oxide and barrium oxide even in the absence of solvent.

Lewis Concept : According to this concept an acid is a substance (molecule or ion) that can accept an electron pair to form a covalent bond and base is a substance that can supply an electron pair to form a covalent bond. Thus an acid is electron pair acceptor and a base is an electron pair donor. An acid need not contain hydrogen.

Lewis acids are of several types:

(i) Compounds having a central atom with incomplete octet.

(ii) Compounds containing multiple bonds.

(iii) Simple cations.

(iv) Compounds in which the octet of the central atom can be expanded.

The acids and bases according to this concept are interrelated by the equation

$$HA + H_2O \rightarrow H_3O^+ + A^-$$

Acid Base Acid Base

Classification of Acids : The acids can be classified as (i) Hydra acids and (ii) Oxy-acids

Hydra acids are those acids in which we find no oxygen, *e.g.* HC1, HBr, HI etc. They contain only two elements *i.e.* hydrogen and some non-metal.

Oxy-acids always contain oxygen as one of the elements. They contain hydrogen, oxygen and a third element. *e.g.* HNO_3, H_2SO_4, H_3PO_4, H_2CO_3 etc.

Relative Strengths of Acids and Bases : The relative strength of an acid and a base depends upon their relative capacity to liberate H^+ and OH^- ions in aqueous solution. The higher the $[H^+]$ in aqueous solution, the greater is the strength of the acid. Similarly greater the $[OH^-]$ is aqueous solution greater is the strength of base.

As already discussed all those substances which give OH^- in aqueous solution are called bases. Out of these *only those bases which are soluble in water are called alkalies.*

Salts : Salts are the compounds formed by the neutralisation reaction between an acid and an alkali.

$$NaOH + HC1 \rightarrow Nacl + H_2O.$$

alkali Acid Salt Water

(base)

NaCl is a neutral salt.

$$NaOH + H_2SO_4 \rightarrow NaHSO_4 + H_2O$$

Base Acid Salt Water

$NaHSO_4$ is an acidic salt because in it hydrogen of the acid has been partly replaced.

Oxidation and Reduction

Oxidation might be defined as a chemical reaction. Wherein oxygen is gained or hydrogen is lost.

Reduction may be defined as a chemical reaction wherein oxygen is lost or hydrogen is gained.

For example when hydrogen is passed over heated Coptic oxide (CuO) the following reaction occurs.

$$CuO\,(s) + H_2\,(g) \rightarrow Cu\,(s) + H_2O(g)$$

CuO loses oxygen and so is reduced to Cu. Hydrogen gains oxygen and is oxidised to H_2O.

The hydrogen which is required to reduce CuO is called *reducing agent* and CuO which is required to oxidise H_2 is called *oxidising agent*.

Definition of oxidation and reduction in terms of electron loss or gain is more useful because all reactions do not involve oxygen and hydrogen. According to this concept.

Oxidation is a process which involves loss of one or more electrons by some atom or group of atoms. For example

$$Cu \rightarrow Cu^{2+} + 2\,e^-.$$

$$Zn \rightarrow Zn^{2+} + 2\,e^-.$$

$$Ag \rightarrow Ag^+ + e^-.$$

$$H \rightarrow H^+ + e^-.$$

The substance which loses electron is said to be oxidised and the one which gains electron is said to be reduced.

Reduction is a process which involves gain of one or more electrons by some atom or group of atoms. For example

$$Cu^{2+} + 2e^- \rightarrow Cu$$

$$H^+ + e^- \rightarrow H$$

$$Ag^+ + e^- \rightarrow Ag.$$

Oxidation and Reduction Occur Simultaneously : We have already studied the electronic concept of oxidation and reduction. If some substance loses electrons (*i.e.* undergoes oxidation) then the electrons lost by it must be accepted by some other sabstance. The substance that accepts electrons undergoes reduction. Hence it is clear that oxidation and reduction occur simultaneously. For example in the reaction.

$$Zn + Cu^{2+} \rightarrow Zn^{2+} + Cu \text{ (redox reaction)}$$

Zn is oxidised to Zn^{2+} and Cu^{2+} is reduced to Cu. The reactions involving simultaneously oxidation and reduction are called *redox-reactions*. A redox reaction can be split into two *half reactions* are representing oxidation and the other representing reduction. For example the above redox reaction may be represented as

$$Zn \rightarrow Zn^{2+} + 2e^- \text{ (oxidation half reaction)}$$

$$Cu^{2+} + 2e^- \rightarrow Cu \text{ (reduction half reaction)}$$

In such reactions the substance that loses electrons is called *reducing agent* and the substance that accepts electrons is called *oxidising agent*.

In a redox reaction the total number of electrons lost by reducing agent is equal to the total number of electrons accepted by the oxidising agent.

Oxidation State : The system of oxidation states (or oxidation numbers) has been devised to give a guide to the extent of oxidation or reduction in a species the system is without direct chemical foundations, but is extremely useful being appropriate to hope ionic and covalently bonded species.

The oxidation state can be defined simply as the number of electrons which must be added to a positive ion to get a neutral atom or removed from a negative ion to get a neutral atom *e.g.* Fe^{2+} (aq) has oxidation state of + 2 and Cl^- has oxidation state of –1.

For covalent species the oxidation state is found using the following rules:

(i) The oxidation state of all elements in uncombined state is taken as zero.

(ii) The algebraic sum of oxidation states of elements in a compound is always zero.

(iii) The algebraic sum of oxidation states of elements in an ion is equal to the charge on the ion.

(iv) The oxidation state of oxygen is – 2 (except in oxygen gas and peroxides).

(v) The oxidation state of hydrogen is +1 (except when combined with group I and II metals as hydrides).

Structure of Atom

John Dalton (1808) proposed that matter is composed of small indivisible particles called atoms.

Particles in an Atom : Atoms are composed *of protons, neutrons* and *electrons*. These are known as fundamental sub-atomic particles. The following table compares the properties of these particles.

Name of particle	*Mass*	*Charge*
Proton, p	1 amu	+ 1
Neutron, n	1 amu	0
Electron, e	negligible	–1

A neutral atom contains equal number of protons and electrons in it. This number of protons or electrons present in an atom is called its *atomic number (Z).*

The total number of neutrons and protons present in an atom gives the *mass number (A)* of the atom.

So

Atomic Number (Z) = Number of protons

= Number of electrons

and

Mass Number (A) = Number of protons + Number of

Neutrons = Number of Necleus

Rutherfords' Atomic Model : According to this model atom consists of two parts (i) nucleus and (ii) extra-nuclear part.

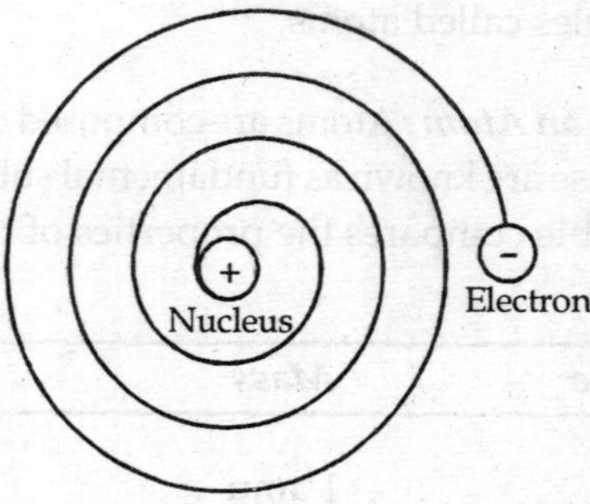

Fig. Gradual decrease in the radius of orbit.

Nucleus. The protons and neutrons in each atom are tightly packed in a positively charged nucleus and the electrons move around the nucleus. Nucleus is a small positively charged part of atom and is situated at the centre and carries almost entire mass of atom. The diameter of nucleus is of the order of $10^{-12} - 10^{-13}$ cm which is only about 1/100,000 part of the diameter of an atom. In chemical reactions nucleus remains unchanged.

Extra-nuclear Space. This is the empty part of the atom. In this part electrons revolve at very high speed in fixed path called *orbits* or *shells*.

Drawbacks of Rutherford's Model. Following serious objections against the Rutherford's model were reported:

(i) When an electron revolves around the nucleus, it will radiate out energy, resulting in the loss of energy. This loss of energy will make the lectron to move slowly and consequently it will be moving in a spiral path and ultimately falling inside the nucleus Fig. Thus, the atom remains unstable. Fortunately, the atom is stable.

(ii) If an electron loses energy continuously, the observed spectrum would be continuous and have broad bonds merging into one another. But most of the atoms give line spectra. Thus Rutherford's model could not explain the origin of spectral lines.

Bohr's Theory. In order to overcome the drawbacks of Rutherford's model and to account for the line spectra of hydrogen, Niel Bohr in 1913 put forward a theory called Bohr's theory. The main postulates of Bohr's theory are as follows:

(a) That within an atom an electron can move in certain specific orbits without radiating out energy. Such orbits were termed as stationary orbits. These orbits are numbered as 1, 2, 3, 4 etc., or K, L, M, N, etc., starting from the nucleus.

(b) The mathematical condition for stationary orbits is that the angular momentum of the moving electron is an integral multiple of $h/2\pi$, where h is the Planck's constant.

$$mvr = n\,(h/2\pi)$$

where mvr denotes the angular momentum and *n* is called principal quantum number and is equal to 1, 2, 3 ....

(c) When an electron gets energy, it will go to higher energy orbits. Similarly in the reverse process, the excited electron jumps down to lower energy level by emitting absorbed energy in the form of radiations of suitable wave length. The frequency of this radiations (v) is given by the difference in the energy between initial and final orbits.

$$E_1 - E_2 = hv$$

Simple representation of sodium atom on Bohr's model.

A sodium atom consists of 11 electrons ($^{23}Na_{11}$) and they are arranged as 2, 8, 1. It may be represented as

Electrons partly because of their very small size are impossible to locate at any particular time. It is however possible to locate a region or volume where the electron is most likely to be found. This region is called *Orbital*. Each orbital can hold a maximum of two electrons. Orbitals can be divided into *s*⁻, *p*⁻, *d*⁻, *f* types. Each type of orbital has its own characteristic shape.

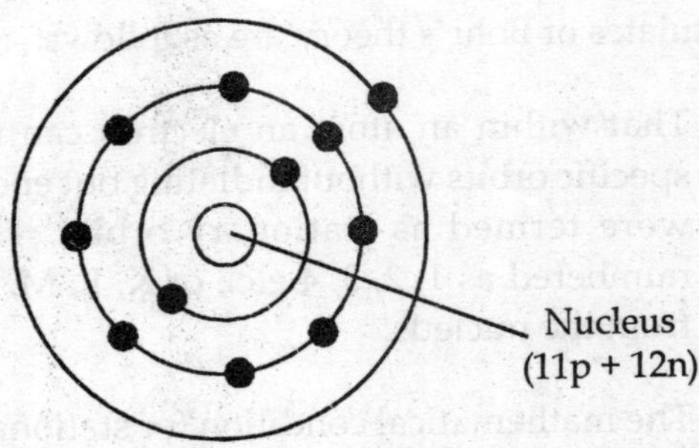

Fig. Simple representation of sodium atom

Quantum Number. The term quantum number is used to identify the various energy levels that are available to an electron in an atom.

Types of Quantum Numbers. Four quantum numbers are necessary to characterise completely any particular electron in a particular orbit. These are as follows:

Principal Quantum Number. This is designated as n and gives the number of principal shell in which the electron is revolving around the nucleus. It designates the average distance of the electron from the nucleus. Hence this quantum number represents the size of electron orbit.

Azimuthal or Subsidiary or Orbital Quantum Number. This is designated as l. This determines the orbital angular momentum and the shape of the orbital. l can have value ranging from 0 to $n - 1$, *i.e.*,

$$l = 0, 1, 2.... (n-2), (n-1)$$

The orbital with $l = 0$ is called s-orbital, that with $l = 1$, is called p-orbital, that one with $l = 2$ is called *p*-orbital and the one with $l = 3$ is called *f*-orbital.

Magnetic Quantum Number. This quantum number is designated as m. This quantum number determines the direction of the orbital relative to the magnetic field in which it is placed. m can have values from – l to + l through zero, *i.e.*,

$$m = +l, l-1, 1-2 .... 0. -1, -2, ... -(l-1), -l.$$

In other words, the number of m values for a given values of l is $2i + 1$ through zero.

Spin Quantum Number. This is designated as *s*. This quantum number indicates the direction in which the electron is spinning clockwise or anticlockwise. There are only two possible values for this quantum number and for electrons these are +1/2 and –1/2 according to the direction of spin being clockwise and anti-clock wise respectively.

Table : Electronic Configuration of the First 20 Elements

At. No.	*Element*	*Orbital Electronic Configuration*			
1.	Hydrogen	$1s^2$			
2.	Helium	$1s^2$			
3.	Lithium	$1s^2$	$2s^1$		
4.	Beryllium	$1s^2$	$2s^2$		
5.	Boron	$1s^2$	$2s^2 p_x^{1}$		
6.	Carbon	$1s^2$	$2_s^2 2p_x^{1} 2p_y^{1}$		
7.	Nitrogen	$1s^2$	$2s^2 2p_x^{1} 2p_y^{1} 2p_z^{1}$		
8.	Oxygen	$1s^2$	$2s^2 2p_x^{2} 2p_y^{1} 2p_z^{1}$		
9.	Flourine	$1s^2$	$2s^2 2p_x^{2} 2p_y^{2} 2p_z^{1}$		
10.	Neon	$1s^2$	$2s^2 2p_x^{2} 2p_y^{1} 2p_z^{2}$		
11.	Sodium	$1s^2$	$2s^2 2p_x^{2} 2p_y^{2} 2p_z^{2}$	$3s^1$	
12.	Magnesium	$1s^2$	$2s^2 2p_x^{2} 2p_y^{2} 2p_z^{2}$	$3s^2$	
13.	Aluminium	$1s^2$	$2s^2 2p_x^{2} 2p_y^{2} 2p_z^{2}$	$3s^2 3p_x^{1}$	
14.	Silicon	$1s^2$	$2s^2 2p_x^{2} 2p_y^{2} 2p_z^{2}$	$3s^2 3p_x^{1} 3p_y^{1}$	
15.	Phosphorous	$1s^2$	$2s^2 2p_x^{2} 2p_y^{2} 2p_z^{2}$	$3s^2 3p_x^{1} 3p_y^{1} 3p_z^{1}$	
16.	Sulphur	$1s^2$	$2s^2 2p_x^{2} 2p_y^{2} 2p_z^{2}$	$3s^2 3p_x^{2} 3p_y^{1} 3p_z^{1}$	
17.	Chlorine	$1s^2$	$2s^2 2p_x^{2} 2p_y^{2} 2p_z^{2}$	$3s^2 3p_x^{2} 3p_y^{2} 3p_z^{1}$	
18.	Argon	$1s^2$	$2s^2 2p_x^{2} 2p_y^{2} 2p_z^{2}$	$3s^2 3p_x^{2} 3p_y^{2} 3p_z^{2}$	
19.	Potassium	$1s^2$	$2s^2 2p_x^{2} 2p_y^{2} 2p_z^{2}$	$3s^2 3p_x^{2} 3p_y^{2} 3p_z^{2}$	$4s^1$
20.	Calcium	$1s^2$	$2s^2 2p_x^{2} 2p_y^{2} 2p_z^{2}$	$3s^2 3p_x^{2} 3p_y^{2} 3p_z^{2}$	$4s^2$

Pauli's Exclusion Principle. This is the most important principle which cannot be derived from any fundamental concept. Pauli's exclusion principle states that no two electrons in a single atom can have all their quantum numbers identical. By this principle it means that if two electrons possess the same value of n, l and m, they must have different values of s.

Hund's Rule of Maximum Multiplicity. This rule has a spectroscopic basis and is mainly concerned with the situation when two orbitals of a sub-group are incompletely filled. This rule can be stated as:

"When electrons enter a set of orbitals in a given shell, electrons will pair up, when all the available orbitals have one electron each". Hund's rule is energetically possible.

Aufbau's Principle. The word Aufbau is a German expression which means build up or construction. This Aufbau principle is mainly concerned with the building up process in which extra electrons are. being added to the various available orbitals so as to balance the nuclear charge. Broadly speaking, this principle states that every electron enters the lowest possible energy state available.

Isotopes. These are the atoms of the same element with same atomic number but different atomic mass (mass number) *e.g.* $^{12}_{6}C$ and $^{13}_{6}C$, $^{1}_{1}H$ and $^{2}_{1}H$.

Isobars. These are the atoms of different elements having different atomic numbers but same atomic mass (mass number) *e.g.* $^{210}_{32}Pb$ and $^{210}_{83}Bi$.

Isotones. These are the atoms having same number of neutrons but different mass numbers *e.g.* $^{30}_{14}Si$, $^{31}_{15}P$, $^{32}_{16}S$. All these have 16 neutrons in their nuclear.

13

Audio-Visual Aids

In an introductory class of audio-visual education from B.Ed. students it was asked some of their earliest school experiences. Following were some of their answers:

(i) "The earliest school experience was when I made things out of plasticine."

(ii) "When in class 3rd we made a model of iglu out of sugar cubes."

(iii) "When I went to see a zoo in my 2nd grade."

(iv) "When we did gardening in a small comer of the school."

(v) "When I played the role of little red riding hood in kindergarten."

(vi) "The puppet show I saw in the school."

Dozens of similar experiences were reported by these (college) students. All these are varied experiences but we see one similarity, they could remember the experiences where they are involved in doing something, when they were active, when more senses were involved. These are some of the experiences where effort was not made to memorise them, on the other hand we do not always remember the facts or concepts that we try hard to memorise. From this we can generalise that learning becomes comparatively permanent when concreteness is there in the experiences. The various experiences help in developing concepts.

The process of building concepts operates naturally from the time a child begins to draw certain conclusions from his experiences and applies these conclusions to new situations. It continues thereafter as he makes new generalisations from new experiences and experiences in which new and old experiences are combined. The overall activity of building concepts, therefore is a realistic definition of Education. Now we can say that two elements are involved in building concepts: (1) Certain amount of concrete experiences, and (2) Combining and recombining these concrete experiences in many ways. When we apply this to classroom teaching, audio-visual aids play a very important role in concept formation and therefore in permanent learning.

Main Features

If we see teaching of science in schools of our country by and large we will find teachers lecturing or even reading out from the books and explaining few things on the blackboard. Some teachers use some demonstrations, charts and models but not very frequently and many a times students memorise things without understanding. Audio-visual aids if properly used help in teaching learning process in many ways (as given below) and can ensure quick and effective learning.

(i) The timely use of proper aids compels attention, develops interest and motivates students.

(ii) They break the monotony of teacher's talk, reduce verbalism, save time (of long verbal explanations) and give a better idea of the real things.

(iii) Audio-visual aids can make learning experiences far more concrete; therefore clarification of concepts, better understanding and long lasting learning are possible.

(iv) Great many teaching problems can be solved partly or wholly by the proper use of rich experiences (through audio-visual aids); therefore they offer great opportunities for improved learning.

The Kinds

Variety of audio-visual aids can be used in science teaching. They can be classified in various ways.

(a) One way of grouping them is as follows:

Visual Aids—charts, photographs, diagrams, static and working models, etc.

Auditory Aids—radio, recordings on tapes, and cassettes.

Audio-visual Aids—films, T.V. etc.

(b) Another way of classifying can be:

Graphic Aids—diagrams, photographs, charts, play cards.

Three-Dimensional Aids—models, specimens, real objects, apparatus, dioramas etc.

Projected Aids—slides, films, etc.

Aids Through Activity—excursions, projects setting and maintenance of aquarium, vivarium, botanical garden etc.

Practical Aspects

Some of these aids are more concrete in nature and some of them are comparatively more abstract. Edger Dale has arranged the various audio-visual aids in pictorial form which he called "Cone of Experiences" as shown in (Fig.).

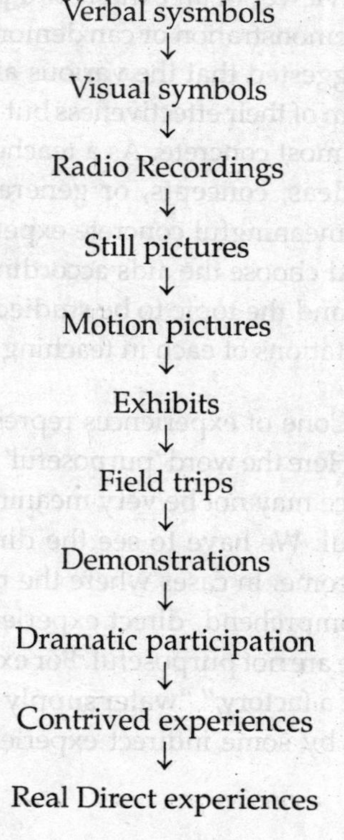

Fig. *Edger Dale Cone of Experiences*

This is a kind of visual aid to visualise and explain the interrelationship of various types of audio-visual material as well as their position in learning process.

At the top of the Cone are verbal symbols which are most abstract and at the base are the direct purposeful experiences which are most concrete form of experiences. This is only a pictorial form where all sorts of aids and experiences are arranged in a Cone. The various bands of the cone representing various experiences and aids, should not be considered as rigid divisions or watertight compartments. They overlap and blend with each other. For example you can be a viewer of an exhibit or a person who made it, you can observe a demonstration or can demonstrate it yourself. It is also not being suggested that the various aids and materials are arranged in the form of their effectiveness but they are arranged from most abstract to most concrete. As a teacher we should also know that abstract ideas, concepts, or generalisations are not possible without rich meaningful concrete experiences. A science teacher has to pick and choose the aids according to the maturity levels of the students and the topic to be studied. We will discuss the strengths and limitations of each in teaching sciences.

The base of the Cone of experiences represents "direct purposeful experiences." Here the word 'purposeful' is very important. Every direct experience may not be very meaningful, therefore, it may not be purposeful. We have to see the direct experience in terms to learning outcome. In cases where the real things are too small or too big to comprehend, direct experiences are not very effective and therefore are not purposeful. For example, "structure of atom," "working of a factory," "water supply in a city" etc., can be understood better by some indirect experiences like models, maps or charts, etc.

In teaching science many direct experiences can be given to the students for effective comprehension. Some examples are: observing real flowers, leaves, plants, insects; dissecting animals;

taking a walk through woods; going to the seashore and observing marine animals; doing salt analysis in the lab; setting and maintaining an aquarium, etc. In such cases learning is by direct participation.

The science teacher has to decide what direct experiences will be purposeful for his classes and then try to give them as many experiences as possible because direct and concrete experiences soon become associated with abstractions and help in developing more difficult concepts.

What are Contrived Experiences? Next in the hierarchy towards abstraction are contrived experiences. Examples of contrived experiences are static models, working models, specimens, dioramas, etc. Contrived experiences may differ from the original in size (big things are made smaller and smaller things are made bigger) and complexity. It usually is a simplified and edited version of the real thing, where the unnecessary details can be removed to make the learning clear. For example a petroleum refinery is difficult to comprehend in a real situation but its model will be more meaningful. In the same way models of places where it is difficult or impossible to reach and see can give a clear idea, like models of globe, volcano, parts and systems of the body, bottom of the ocean, view of a forest, polar region, etc. Sometimes we imitate the whole natural habitat in the form of dioramas and keep in museums. Cut-away or half cut models are extremely useful in teaching internal structure of eye, ear, stem, root and automobile, etc.

There are some examples where nobody has seen the real thing, the models are made on the basis of indirect evidences like model of atom, DNA structure. Here the whole concept is developed on the basis of these imaginary models.

While teaching through the models, teacher should give the idea of the real thing, regarding their size and complexity. All models are not correct reproduction of their originals, they are only simplified versions.

Objects and Specimens. Objects and specimens are very common in science laboratories. These are also examples of contrived experiences. We collect rocks from various places. Different kinds of plants and animals are collected; pressing, preserving and sniffing are done for storing and study purposes. The objects and specimens are taken from the real settings. They are samples of real things minus real settings. The specimens are collected and stored so that they are readily available for study purposes. Another most important advantage of objects and specimens which is not otherwise possible in direct experience is that they can be arranged into groups and classes.

Various Types

This has been placed on the 3rd band of Cone of experiences. Dramatisation means substitute for real experience of reconstruction of the original reality. There are many things we cannot possibly experience at first hand. There is a great value of dramatisation in education. Students can participate in a dramatisation or watch some kind of dramatisation. Both are valuable experiences but participation is much more meaningful and closer to reality than only watching.

The question is "What is the scope of dramatisation in science teaching?" Dramatic acts are quite popular in languages and social sciences. In sciences also the scope is not limited.

The films made on the work of various scientists are only possible because of dramatisation. Such films are quite effective as such experiences are otherwise not possible.

Students of primary and middle classes participate in scientific dramatised act in schools, such activities are also brought to science fairs etc. Some very abstract and uninteresting ideas are taken for dramatisations, for example, different students act as various components of solar system with proper costumes, dialogues, songs, music and dance, and the abstract concepts become clear and leave a long lasting impact on participants and viewers; students act as various petroleum products and explain how they are formed and utilised like coal, petrol, vaseline, synthetic rubber and plastic etc. Functions of vitamins and other components of food can also be taught through dramatisation.

It is also possible to use dramatisation in classroom teaching where costumes are not necessarily required but different students can remember their parts and act out in the classroom.

Interdependence of various components of an ecosystem and balance of nature can also be explained interestingly through dramatisation. An imaginative science teacher can think of many such topics which can be taught more effectively through such activities. Thus dramatisation can become an effective teaching aid in teaching science.

Demonstrations : Demonstrations are quite familiar activities in science class-rooms. Their merits and demerits have been discussed at length in the Cone of experiences they have been placed on the fourth band from the base because it is essentially a process of observing. It differs from the first three bands, which are essentially doing. Demonstrations are used to show how something is done or not done. When demonstration is followed by doing on the part of students, it becomes very much meaningful.

Demonstrations are used to clarify ideas, and help to develop skills, processes and attitudes. They are not limited only to demonstrations through apparatus only. They can be used to clarify abstract ideas on the chalkboard, through slides or motion picture also.

Demonstrations can be improved to a great extent if certain points are kept in mind while planning and doing a demonstration.

- Plan all steps of demonstration in advance.
- Rehearse the demonstration before going to the class-room.
- Keep the demonstration simple as far as possible.
- Keep the students involved.
- Make it sure that all the students can see it.
- Outline various points on the board.
- Keep summarising various steps.

What is Field Trip or Excursion? A field trip or excursion is a planned visit to a point outside the regular classroom. It may be in the school, out in the community or it could be a long trip to far away places. Usually in field trips to places like visits to a factory, observatory, agricultural institute, poultry farms, museums etc., we often see other people doing things. As spectators we are not involved but we directly watch it and get a first hand knowledge. Therefore field trip is an excellent bridge between the work of the classroom and the work of the outside world. The chief difference between a field trip and other educational experiences is that the student get their experiences in the field and not in the classroom. Because of its nature to be mostly observation it is kept on the 5th band of the Cone of experiences as less concrete than other experiences discussed before. But if the field trips are planned and arranged in such a way that they go beyond observation, for example of a sea-beach, on a pond where they can also touch, feel and collect things, it becomes a *direct experience* for them. Such a variation in the field trip indicates again how the bands of the Cone interlap and blend into one another.

Importance of the Field Trips. Excursions or Field Trips are of great educational value especially in science subjects. The classroom is a limited place, bounded normally by four walls and meagrely equipped for the task or providing students with worthwhile experiences. The environment outside the classroom has no bounds; it has almost every conceivable situation that a teacher might wish to utilise. In school corridor students may study writing system for supply of electricity to different rooms and laboratories, may determine the power (in watts). On school grounds there may be various types of plants, birds, insects, different kinds of soil, sun-shine and shadows, building materials, bicycles, scooters and cars. And just beyond the school boundaries lie the unlimited resources of community.

Contributions of an Excursion or Field Trip

(i) Field experiences are first hand experiences. They arise from direct learning situations. Sometimes they play the same role or even better in the learning of science as do experiments and demonstrations.

(ii) Field experiences tend to be much more meaningful and permit easier transfer of learning to solutions of real life problems.

(iii) Fieldwork if properly organised awakens many interests that classroom work cannot arouse. Fieldwork is the study of actual objects which stimulate more curiosity than to ideas. Out of almost any situation encountered in the field can develop into some challenging problems.

(iv) Fieldwork permits first-hand study of many things that cannot be brought into the classroom because of size and other inconvenience, *e.g.*, it is only outside the class that the students can be acquainted with the flora and fauna of the area.

(v) Fieldwork permits a class to engage in activities that are too noisy or too violent to be used in the classroom, *e.g.*, a model airplane, gasoline engine if demonstrated in the classroom would disturb the other neighbouring classes too.

(vi) Outdoors, students are able to work with large size materials, *e.g.*, an iron piece pipe makes a more impressive lever than a routine stick and erosion is better demonstrated from a garden hose than with a tiny trickle from a yet drawn in a glass tube.

Some Special Experiences of Field Trips. Fieldwork brings students in contact with many objects. The observation and manipulation of objects in the environment are bound to arise questions. Attempts to answer questions give rise to new problems.

Field trips can be used for review and drill. Ideas learnt in classroom can be better fixed in mind in actual situations, *e.g.*, by visiting factories the ideas of running a plant and the products manufactured can be better fixed in students' minds, or visiting a zoological park or a botanical garden, more abstract ideas about animal and plant life can be more clarified to students.

Difficulties Experienced in Arranging Field Trips. Lack of availability of good inventories regarding field experiences, school policies, and transportation problems create hurdles in arranging field trips. Therefore the teacher will have to take a key role if he wants to arrange the trip.

Procedure to Arrange Field Trips

1. Survey of the place of excursion should be taken before going to the field trip, so that the teacher knows beforehand what their students are to see there; or what relevant literature should be studied if places are far away.

2. Objectives of the field trips should be very carefully identified to make a field rip a success. If this is done teacher knows what he is going to teach and the students know what they are going to learn there.

3. Permission from the authorities and parents of students should be taken well in advance.

4. Appropriate activities compatible to the identified objectives snould be listed and given as well as discussed with the students beforehand.

5. If there is a need for transport and place to stay they should be arranged in advance.

Preparation of Students/or Excursion

1. It is essential to brief students after arriving at the area. This should be as minimum as possible. A long lecture may deprive the students of an opportunity to explore the place.

2. Students should be divided into small groups appointing a group leader for each group. Group leaders may help in running fieldwork smoothly and quickly.

3. You as a teacher should be aware of the fact that there are slow as well as rapid learners. Activities and responsibilities should be so divided among the students as to be equally shared by both type of students without any feeling that they are slow learners or rapid learners.

4. Students should be made aware that there may be some difficulties like: (a) noise while visiting any working plant in a factory, (b) listening to the guide due to distance while standing around him, or (c) technical language used in explaining particular process; and they

should try to solve such difficulties themselves as far as possible.

5. Students may be advised to take notes and draw diagrams, whenever they think it is necessary.

Teacher's Role. The teacher's role is very important for a successful field trip. Some of them are listed below:

1. Watches students closely and gives specific help as and when needed.

2. Recognises student's achievement.

3. Avoids frequent interruption in student's work but occasionally if it is justified.

4. Does not lecture while in field ?

5. Avoids loud voiced comments.

6. Acts as a guide, resource person or consultant.

7. Does the relevant follow-up activities of the trip ?

Follow-up Activities. Follow-up activities are very important in any kind of field trip, it may be a short trip just outside the classroom or a long excursion to various far away places.

Follow-up activities vary according to the nature of excursion. If the field trip is taken just outside the classroom in the school lawns to find the population densities of various species of plants, the follow-up activity will include the pooling up of data collected by each student or each group of students and discussion on the result. If the field trip is an observational trip to places like factories, mills, observatories or hospitals, discussion about what they have learned is important. Field trip can also be taken to study the fauna and flora and collection of specimens from seashore, hill stations or any other places. If such field trip is taken by the students follow-up activities will include pressing of plants, preservation of animals, drying of insects, classification and displaying of material. If proper care has not been taken just after or during the trip (in case of long trips) the materials collected cannot be utilised properly and goes to waste. Sometimes the students and teachers are quite enthusiastic while planning and taking the field trip, but forget about the material after the big excitement or do not do the follow-up activities properly, therefore, the field trip is not as effective as it should be.

In short we can say that follow-up activities should be planned according to the objectives of the field trip. Science teachers should plan their objectives for the field trip well and do the follow-up accordingly for better understanding.

Exhibits : There is a big variety of exhibits. Sometimes they are three-dimensional working models, sometimes a series of photographs or photographs mixed with charts, models and real objects. But exhibits are essentially something one sees as a spectator. Usually one is not involved in handling any thing or working with the material. Science teachers take their students to show the science fairs where different kinds of exhibits can be seen. They are also of great educational value. Sometimes they may also influence the attitudes.

When students are involved in making exhibits, it becomes a direct experience for them. Investigatory science projects done by students can be displayed as exhibits. Exhibits can also be used by the science teachers to teach subject matter.

Museum. Museums are places where items and exhibits of knowledge are assembled, protected and studied. In big cities there are public museums of various kinds like art museums, history museums, science museums, and natural history museums etc. Here we are concerned with science and natural history museums. In our country there are not very many science and natural history museums. But if they are available in your cities, advantage should be taken from them.

Mostly Science and Natural History Museums (Science museum in Bangalore, Natural History Museum in Darjeeling and New Delhi) have two functions: (i) presentation and display of the materials, and (ii) to work with classroom teachers on specific curriculum unit. If this facility is given by the public museums, science teachers should take full advantage of such arrangements.

The other kind of museums are college and school museums for various subjects. Biology museums are quite common museums established in some form or the other in many schools and colleges. Museum should not be considered as a collection of various items and objects. It is more than that, it must be viewed with an idea, a process, and an objective. Therefore science museums should not emphasise only on the product of science but due importance should also be given to science processes.

There is a big scope of displaying of specimens and objects collected by students. Models and projects made by students could also be displayed there. In fact development of museum can be taken as one of the science club activities.

Motion Pictures and Television : Next on the cone towards little more abstraction comes the TV and motion pictures because we are only the viewers of these audio-visual aids. First we will discuss the motion pictures.

Motion Pictures 16-mm Educational Films. It is not possible to learn everything by doing or first-hand viewing. We have to get some of our experiences indirectly, 16-mm films play a very important role in giving this indirect experience. Motion pictures present an abstracted version of the real events omitting unnecessary and unimportant details, they can, sometimes dramatised events so effectively filmed that we feel as though we are present at the reality itself. No other medium has brought so much information and scientific knowledge to the classrooms as the educational sound motion pictures.

There are various kinds of motion pictures. Here we are concerned with the *educational films.* Educational films include *documentary* and *instructional films.* Though some documentary films can be used in classroom teaching but they are especially planned for classroom teaching purposes. On the other hand *instructional films* are specially planned to achieve certain educational objectives and are made in specific subject areas tor teaching purposes. These films can be background films or direct teaching films. They help to promote to achieve a skill, an attitude or to convey certain facts, information, phenomenon or theory. As a science teacher we will be mainly concerned with the instructional films in our areas.

Though we are only spectators before a motion picture but the learning outcomes are quite effective because of certain specific values of motion pictures.

For 16-mm films, 16-mm movie projectors are needed. Now video cassettes of 16-mm films are available, which can be seen on TV. with V.C.P. or V.C.R.

Strengths of Motion Pictures

(a) Motion pictures motivate the students and compel attention.

(b) Movement can very effectively be shown by motion pictures which is not possible in any other aid except direct experience.

(c) It is an edited version of reality, the editing involves manipulation of time, space and by eliminating distractions; it gives relationships of things, ideas and events that might well be overlooked in real life. For example, food web, food chain, etc., can be seen as continuous processes which is not possible in reality.

(d) Time can be controlled in a motion picture with the help of slow motion photography, "movements in animals, insects, flight of birds" can be understood very easily. With the help of time-lapse photography it is possible to see the blooming of flowers, germination of seeds, growths of plants, etc.

(e) Motion pictures can enlarge or reduce the actual size of the objects for better compensation. By means of microscopic lens attachments microscopic organisms like algae, fungi, protozoas, and blood circulation in capillaries etc. can easily be seen and understood by students.

(f) By X-ray cine photography techniques movements of internal organs of animals—action of arms, legs, movement of the heel, swallowing of food, heart-beating/living movements, etc., can be shown.

(g) Motion pictures can bring past and distant present into the classroom.

(h) With the help of animation techniques motion pictures can show processes which cannot be seen by human eye even with the help of microscope or telescope. For example, behaviour of molecules in solids, liquids and gases, phenomenon of nuclear fission and fusion, etc.

(i) Motion pictures can reproduce the record of events like lunar and solar eclipses, etc. which are not always possible to see. Complicated techniques can be filmed and reproduced when required.

(j) Motion pictures promote understanding of abstract relationship, offer a satisfying aesthetic experience, influence and even change in attitudes.

With so many values, motion pictures not only expedite the rate of learning but they also increase the scope of teaching. This is an important consideration in the field of science where mass of learning and the learning processes are expanding so rapidly that it is to keep hand up with them. Motion pictures enable the teacher to clarify the difficult concepts in less time than with any other technique.

In order to get the most effective use of films or video tapes, the following necessary information and skills should be acquired:

(i) Where to obtain appropriate films or video tapes?

(ii) How to use such films and video tapes in your teaching?

(iii) How to operate the 16-mm film projector, V.C.R, V.C.R.?

School, college and university teachers have to spend some time to find films or video cassettes which they can integrate into their courses. It is possible to find suitable films or video cassettes on various science subjects. Securing catalogues will be important for this purpose.

For an effective use of films or video cassettes, preparation and follow-up are two very important steps to be taken by a teacher. Teacher should select the film or video cassette which goes along with the topic. The film or video cassette should be previewed and important points to be covered in the lesson should be outlined.

It will be advisable to frame few short answer questions. These questions can be given to the students before showing the film or video. During the follow-up activity unanswered questions and other important points should be discussed. Such activities help in coordinating the film or video with the topic. In the absence of such coordination sometimes the film or video becomes a wastage of time and effort.

The next point is to learn to operate the 16-mm projector and V.C.P./V.C.R. which is not very difficult to learn. Ordinarily only a few hours instruction followed by a brief practice results in satisfactory operation. Simple threading and operating instructions are given somewhere on the projector, which should be followed carefully. The possibilities of error should not discourage the teacher but it should be taken as challenge. The thrill of this educational device will compensate for minor disappointments.

Caution to be Observed in Using Films or Videos. Motion pictures may not necessarily be effective for teaching everything. As a teacher you have to be cautious about certain things.

- You have to think about the *effectiveness* of the film or video. If other experiences like direct or contrived are more meaningful, film or video should not be used.
- Sometimes the film or video gives a wrong idea about the *time notion* and *size notion*. It should be clarified to the students by comparing it with a known object or by giving a familiar example.
- If inexpensive devices are available with the *same effectiveness* as motion pictures, then less expensive devices should be used.
- Teacher should see that the film or video is adequate for the comprehension of his/her students.

School T.V. Programmes. Television has all the potentialities of motion picture with a little more concreteness because of the "on the spot coverage" and the nature of the TV programmes, as most of them are specially produced for a particular audience. This is more so for educational TV programmes. In many countries educational TV programmes include series of science programmes.

Educational TV in India. Experimental TV Service was inaugurated on September 15th, 1959. This programme was planned for educational and cultural values of community and designed for community viewing. In the meantime Ford Foundation in India was approached to assist in the development of educational TV. Ford Foundation team of TV experts visited India in February 1960. In 1961 Television was started in India as an educational TV project. The project was launched in close collaboration with the Delhi Directorate of Education and with the financial assistance of the Ford Foundation in 1961.

The educational TV programmes were integrated with the school syllabus based on the principle of direct teaching. Direct teaching implied that the TV programmes were directly related with the school syllabus. They were proposed to help in classroom teaching by bringing into classroom additional resources through the medium of TV. The students were benefited from talented teachers, best and expensive equipment and other audio-visual aids which were not usually available in schools. In the project the teachers were also fully involved from the planning to evaluation. This arrangement of instructional TV was working well in Delhi schools because there was a set prescribed syllabus. In some other advanced countries there is not any set syllabus for all but they have different syllabi choosen by teachers and students. Their educational TV programmes are not strictly integrated with the syllabi.

In Delhi the educational programmes started in General Science, Social Studies, Physics and Chemistry. Later Biology Maths and Geography were also included.

For proper functioning of educational TV and better coordination between Doordarshan and Schools a Television Branch was set up in 1967 in Delhi Directorate of Education. The TV branch used to divide the syllabus term-wise and week-wise, and select lessons for TV teaching. To ensure close relationship between TV and classroom teaching, a package of a uniform time-table, set examination days and working hours, and a uniform weekly syllabus in the form of a TV booklet was to be distributed to all viewing schools in the beginning of the school session. Such TV booklets used to be prepared for each subject taught on TV. But now we do not have such school TV programmes in Delhi or elsewhere in the country.

This kind of educational TV programme if started again can solve several problems of science teaching in our schools.

Merits

1. It solves the problem of unequipped laboratories to certain extent. Students are able to see effective demonstrations and expensive apparatus which is otherwise not possible for them to see.
2. TV teachers prepare and rehearse a lesson, spend enough time to collect relevant material and up-to-date information.
3. It has all advantages of films plus being a live presentation seems more natural. Only relevant portions of films are shown, along with teacher's explanation and other visual material.

Demerits

1. It is a one way process and students are passive viewers.
2. There is no opportunity for an active participation of the learner during TV lesson.

Better Utilisation of Educational TV. The disadvantages can be removed if some preparation is done and precautions are taken by the classroom teacher.

For effective use of any mass media three things are important: proper reception, proper seating arrangement, carefully planned preparation and follow-up.

Good reception is one of the basic important things for effective viewing. The television branch in Delhi had the responsibility of installing and maintenance of TV sets. The coordination had to be done by the teachers and administration of the school. Classroom teacher was also be familiar with proper handling and minor adjustments of a TV set.

Next important thing is the proper seating arrangements. If a big crowd is sitting on the floor (without any discipline) in an auditorium or classroom, not sure of proper viewing, this kind of arrangement will be a waste of time. The classroom teacher was to see that there was proper seating arrangement.

TV presentation is only one part of the learning procedure. Preparation for the TV lesson or the pre-telecast activities and the follow-up or post-telecast activities by the classroom teacher are equally important without which the TV lesson may not be very much meaningful.

Pre-telecast activities included the preparation by the subject teacher. Preparation varied with class to class and also depended on the topic, the classroom teacher could see the topic of the TV lesson in advance from the TV booklet. During pre-telecast activities his job was to make the students ready to receive the TV instructions. He could ask motivational questions. If some background knowledge was needed it was to be given at this time. The time-table was arranged in such a way that there were about 10 minutes before and after the actual telecast, usually the presentation was of 20 minutes duration, and class periods varied from 35-40 minutes.

After the TV presentation follow-up activities included clarification of points of doubts raised by students. Recapitulation questions could also be asked to see how much they have grasped. With little encouragement by the classroom teacher many activities could be followed by students as continuation of the TV presentation for better understanding.

The TV branch of Delhi also tried to evaluate the lessons. They used to send a printed pad of evaluation check sheets to each school for the constant feedback of TV lessons. If various people connected with educational TV production do their jobs consciously the E.T.V. programmes can be quite effective. You as a science teacher has to play an important role. During your practice teaching you can observe some science lessons on video tapes, evaluate them according to the evaluation check sheet developed by you and give suggestions for further improvement.

Radio Recordings and Still Pictures : Next stage on the cone of experiences after motion pictures is the number of those devices which can be called as one-dimensional aids. We will take the scope of each in teaching science one by one.

Kinds of Still Pictures : Various kinds of still pictures can be used in teaching science. They include photographs, illustrations and slides. These visual materials are of great importance in teaching learning situations.

Photographs and illustrations can be used without projection, slides can be projected on the screen with slide projectors. Therefore pictures can be divided into two kinds: (i) unprojected, and (ii) projected.

Unprojected Pictures : Photographs, illustrations and clippings of photographs from magazines and newspapers can be collected for teaching purposes. Pictures give a correct impression of the object or situation and motivate and enrich teaching. It is good to have relevant pictures on the walls of science rooms—

pictures of great scientists and some of their apparatus, etc. Pictures can also be used in teaching for motivation, introduction, presentation or recapitulation as the need may be.

Pictures and illustrations with some theme can be displayed on a bulletin board.

Bulletin Board : It is a place for posting "bulletins", as the name suggests. Bulletin boards can be used in many different ways and as an effective "teaching device". Bulletin boards inside and outside the science rooms and laboratories can be used for putting visual materials (pictures, illustrations) with relevant heading and some labelling for supplementing the teaching. It should be of scientific interest and importance. With proper teacher guidance and motivation bulletin boards can provide opportunity of developing creativity, responsibility and interest among the students. Other than displaying the photographs and illustrations, bulletin boards can also be used for displaying reports of science projects done by students, science news and cuttings from science magazines, collected and presented by them.

Projected Pictures : Projection of visuals have certain advantages. For example, projection magnifies the material, therefore, a bigger group can easily see it. It is somehow motivating, therefore, compels attention.

Different kinds of projectors aroused in classrooms for projecting still pictures. They all are very easy to use, with little planning on the part of the teacher, teaching can be quite effective.

Opaque Projector of Episcope : The opaque projector projects and simultaneously enlarges material directly from original books, magazines, etc. All kinds of written, printed or pictorial matter in any sequence can be projected for teaching purpose. It can also project opaque thin objects like leaves, shells, sample of fabrics, butterflies, moths, etc.

Designs for posters, maps, pictures for displaying on bulletin boards etc., can be projected in the desired enlarged forms and then traced. The opaque projector though has many unexplored possibilities but is not being used much because of its bulk, weight and requirement of a very dark room.

Slide Projector : Such projectors usually project 35 mm slides. It is essentially a simple machine, easy to operate, inexpensive and light weight. It has most of the points of practical importance.

Slides can be shown individually in any desired order.

(i) Slides are of great value in visual teaching situation, when motion is of little or no importance, for example, in teaching structural details.

(ii) Great variety of visual material such as pictures, charts, graphs, diagrams, maps, tables, anything that can be photographed can be put on a slide.

(iii) Slides are available in black and white and coloured forms on many science subjects covering a wide range of topics and levels.

(iv) Slides are comparatively inexpensive.

(v) Slides require only a slight darkening of the room.

(vi) In some little more expensive projectors the slides can be moved automatically either from a master control or from a synchronising device on a sound track. This arrangement of slide-tape programme is an effective device as the well prepared commentary makes it more meaningful.

The use of this visual aid will be more meaningful if following points are considered while using them:

(i) The science teacher has to explore if suitable slides are available on the particular topics.

(ii) Second step is the careful preview. The preview helps in deciding on how to use the material to the best advantage. Is it stimulating enough for introduction on the whole lesson could be developed with it or only suited for summarising the lesson.

(iii) Teacher has to think about the purpose also. If showing of motion is necessary for understanding, then slide is not the best medium.

(iv) If teaching involves a series of step by step development, one leading to the next logical sequence the slides are well suited.

Showing slides is only one part of the process. Good discussion and further activities could be developed from it.

The Micro projector : One of the more specialised types of projection equipment is the micro projector. As its name indicates it shows enlarged images of stained sections of microscopic slides or other biological material mounted on microscopic slides so that the bigger group can see it.

The advantages of micro projector are that:

(i) It presents an enlarged picture of the object on the slide for common view.

(ii) It assures the teacher that his students are seeing precisely what he wants them to see.

Some manufacturers make low power projection attachments for slide projectors. Some are simple extensions of a microscope using a more powerful illumination system with a small view screen. This can be easily shown in small groups. But the full scale micro-projectors are quite expensive.

Projectors of any kind are a boon to classroom teachers. These technological tools, if coordinated wisely make the classroom teaching quite effective, because of the novelty of presentation, students become motivated and interested.

Radio and Recordings

Language and Voice Modulation. These are primarily auditory aids. If the speaker on radio or recordings uses simple everyday language with meaningful associations with familiar examples the listening can become quite concrete. On the other hand if the language is difficult without any associations, it could be an abstract thing for listeners. Being only an auditory aid language and voice modulation etc, play a very important role in the effectiveness of these programmes

Values of Radio. Immediately we are able to get the news of important events from all over the world, which is not possible from any other media except vedio or TV. Many events are of great importance for science classes, some of which are mentioned below :

(i) Radio brings reality to the classroom The spectator who describes the events with his tone and description can brings sense of participation amongst listeners.

(ii) Radio programmes if well rehearsed and well presented in the form of stories bring the personal feeling of the actor, different sound effects and voices make them more realistic and meaningful.

(iii) It brings variety to the classroom and helps in teaching

(iv) Though in a well prepared programme enough expenses may be involved but radio being a mass media technology, on the whole it is inexpensive.

Limitations

(i) It is still a one way communication

(ii) Availability and maintenance of radio set may be a problem.

(iii) Radio broadcast timings always do not coincide with the class timings. Provision of radio broadcasts in the school time table is made only in a very few schools.

Educational Radio Programmes in Our Country. Educational broadcasts for schools in the AIR were regular features from the last few decades. These programmes included science lessons for students and teachers at various levels. Teachers' programmes were on content as well as on methodology. Before the board examinations revision lessons on various science subjects were quite popular. These were used to be given by experienced science teachers and educators

The school broadcast was a separate unit of AIR in New Delhi. The unit tried to coordinate between various agencies to improve the effectiveness of the broadcasts. Meetings were held where people were invited from Directorate of Education Delhi, Municipal Corporation Delhi, New Delhi Municipal Council and Delhi University to plan future programmes and give suggestions for their improvement

Radio broadcast schedule in the form of a chart was prmted and sent to the schools in the beginning of the year. Teachers could utilise the relevant broadcast in their classes. Radios and transistors were or could be easily available in the schools. If the teachers are conscious of the radio programmes students can be benefited Pre- and post-broadcast activities improved the effectiveness of the programmes as discussed in TV programmes.

Recordings. All the values of radio can be applied to recordings also. But the disadvantages can be eliminated like timing, administration and one way communication problems. Recordings from many sources can be utilised in the classrooms, like spool-tape records and cassette records.

Advantages of Recordings

(i) Recordings can become two way communication as it can be stopped, discussed and replayed wherever it is necessary, it can be controlled by us.

(ii) Recordings on cassette or spool-tape-recorder can be made in the school also. A commentary can be written and recorded on the available slides in your subject in the school for effective use and re-use.

(iii) Good radio programmes can be recorded and replayed in class whenever needed.

(iv) Scientific talks by eminent personalities can be taped and reproduced in the class.

Symbols in Use

Kinds of Visual Symbols. They include the abstract representation of real pictures. Various kinds of visual symbols used in communication are drawings, sketches, diagrams, graphs, cartoons, flat maps, etc. They all can be projected on screen, and can also be shown on the blackboard or on the overhead projector.

Diagrams and charts etc., are part of everyday work for most of the science teachers. Almost all the science teachers use blackboard to show in the structure and happenings of various things in their subjects.

Commercially made charts are available in various subjects. Charts and other graphic aids are also made by various

Governmental agencies. Central Institute of Educational Technology (CIET) of NCERT New Delhi develops various kinds of graphic aids for school purposes. Charts are also made in schools by teachers and students and can be utilised in teaching science quite effectively.

Chalkboard. Blackboard now is usually called a chalkboard because of variations in colours. It is one of the very important but quite neglected aid of the classrooms. If it is properly used can become a valuable aid in teaching science.

It should not be taken for granted that you already know the use of chalk board, but the techniques should be learned and practised especially as a trainee teacher.

Importance of Chalkboard

(i) This aid is used with all other aids for better clarification, for example, the main points of a demonstration planning and summarising of a field trip, questions for a motion picture and so on can be written on the board.

(ii) It gives a concrete form to abstract and vague statements and ideas.

(iii) It helps in developing the skill of drawing a particular diagram..

(iv) It is less time consuming and comparatively cheaper.

(v) It can be used for some students activities like quizes, competitions and discussions, etc.

(vi) Some chalkboards can be used for permanent outlines of maps, graphs, tables and other special materials where some filling up can serve the purpose.

(vii) If properly used chalkboard becomes an attractive point and holds students attention.

Effective Use of Chalkboard

(i) Clean the board properly before using it.

(ii) Try to write neat and bold so that it is legible.

(iii) Do not write too much on the board. Start writing from top left hand. Give headings and sub-headings in the written summary.

(iv) If diagrams are used for referring or introduction, they can be made beforehand. If you want the students to draw the diagrams then it should be drawn in front of them and they should be asked to draw side by side. In the same way a complicated process should not be drawn beforehand, but it should be developed before them for clarification and better understanding.

(v) If necessary, some tools can be used to draw on the board.

(vi) Coloured chalks should be used for making complex diagrams.

Overhead Projector. Any thing which is written or drawn on a chalkboard can also be projected by an overhead projector with some additional advantages.

An overhead projector projects material overhead or over shoulder the teacher using it on to a screen behind and above him. They generally have 25 cm x 25 cm writing area which is big enough for writing. What is required to be shown on the screen is written on a glass plate covered with sheet or roll of transparent plastic. Variety of fabrics tipped pens or china marking pencils are used for writing and drawing. Wet cloth can be used for erasing and re-using the sheets.

Advantages of Overhead Projector

(i) It is very simple and convenient to use the machine.

(ii) This projector unlike any other projectors is operated from the front of the room and the teacher faces class as he or she writes or points on already made transparencies.

(iii) Plastic roll attachment is quite useful. Diagrams, points of the lesson, assignments, tests and similar material can be effectively presented gradually with a minimum of time and effort.

(iv) An important feature of overhead projector is the provision of "overlays" or 'build ups," which have successive layers of transparencies in black and white coloured or both showing cumulative stages of development, sequences and sectional views, which is quite effective in many scientific processes and phenomenon.

Limitations

(i) Though it is not very heavy but a voluminous projector.

(ii) It is quite expensive and needs some planning.

This is the most abstract and one of the most important forms of experience we get and give to our students. Our job as a teacher is to make sure that the various terms, ideas, words, principles and other abstractions used should be meaningful for students. The ideas and concepts which can be translated into verbal symbols come by getting other concrete experiences. Rich experiences gained by students help them to understand the verbal symbols meaningfully and therefore they can read and understand the textbooks also. Ability to speak or read the word does not mean that one also understands the object, process or phenomenon.

If from the very beginning science is taught by reading books or abstraction only, students usually acquire the habit of memorising and accepting verbal formulation given by others. This is no way of teaching science.

Various kind of experiences discussed on the core of experience from direct purposeful experience to visual symbols, all help in the development of verbal symbols meaningfully. Then reading is not only the process of reproduction of verbal symbols but also involves a thinking process, which includes meaning of words.

Textbooks and other reference books provide the most to the students, because all teaching is not possible by any one particular aid or experience. But the textbooks should be written in understandable vocabulary of that level. Sometimes in primary classes the language of science textbooks is much more difficult than the language in the language textbooks of that particular class. Another important aspect is also the style of writing and format. In writing and specially in teaching if references and examples are given from their environment and known things learning will be more meaningful.

Effective Use

A variety of teaching aids has been discussed in this chapter with some details. It is difficult to select few aids and to say that they are more important than others, or mis should be used more often than others etc. The value of an aid does not depend wholly on its quality itself but it also greatly depends upon the way it is used and the particular time it is used.

The qualities of various aids have also been discussed. No matter how perfect the aid may be, if it is irrelevant to the occasion of its use then it is of little value and an ordinary aid with not too many qualities becomes indispensable if it is used relevantly at the right time. For example, there is film which shows the animals under the sea or life in a desert in relation to their ecology. Neither the materials nor this relationship could be shown in a class by any other media or field trip as effectively as the film could. But always it should not be used as a substitute for other materials of the classroom.

QUESTIONS

1. What is the importance of audio-visual aids in teaching science?
2. What are various kinds of audio-visual aids which can be used in teaching science?
3. How will you use the following activities or aids effectively in your teaching? Explain with examples.
   - (i) Direct experiences
   - (ii) Contrived experiences
   - (iii) Demonstration
   - (iv) Science museum
   - (v) Still pictures.
4. Identify some places where field trips can be arranged for meaningful learning in your teaching subject. Explain how will you arrange and carry out this activity.

5. What is the importance of 16 mm films? Identify five topics in your teaching subject where films can be effectively used.

6. Discuss the role of the following in teaching science:

   (i) Slide projector

   (ii) Overhead projector

   (iii) Micro projector

   (iv) Opaque projector.

7. As a science teacher how will you utilise educational TV for your classes?

8. How will you use the bulletin boards in your laboratory or classroom?

9. What precautions should be taken while selecting audio-visual aids in your teaching?

10. What audio-visual aids would you like to have in your science laboratory? Justify your answer.

14

Development of Curriculum

Dissatisfaction with the existing curriculum is natural in a keen and up-to-date teacher of any subject, particularly if, like chemistry, that subject is itself undergoing change. Such dissatisfaction provides the impulse for reform of science curriculum within schools and leads, usually gradually, to changes in both content and teaching strategy. During 1950's considerable amounts of money were made available in several countries for large scale reforms. Large scale curriculum development started in the united states in 1950's and were taken up in Britain in the 1960's. During 1960's curriculum reforms were initiated in many countries all over the world. It would not be an exaggeration to say that the changes in school chemistry that have occurred on a world-wide scale during the 1960's and 1970's have greatly exceeded those of the previous fifty years.

The Concept

Curriculum is a gist of lessons and topics which are expected to the covered in a specified period of time in any class. However this traditional concept of curriculum has undergone a change in modern times. Now curriculum refers to the totality of experiences that a child receives through various class room activities as also from activities in library, laboratory, work shop, assembly hall, play fields etc. Thus according to modern concept curriculum includes the whole life of the school. Thus those activities which were previously referred to as co-curricular or extra-curricular activities have now become curricular activities.

According to this concept the curriculum can be considered to include the subject matter, various co-curricular activities etc.

Curriculum is derived from Latin word "currere" meaning "to run". Thus curriculum in the medium to realise the goals and objectives of teaching a particular course of study.

Basic Principles

There are certain basic principles of curriculum planning which should form the basis for the formation of a good curriculum. These are:

1. *The principle of child centredness:* The curriculum should be based on the present needs and circumstances of the child.
2. Curriculum should provide a fulness of experience for children.
3. The curriculum should be dynamic and not static.
4. It should be related to every day life.

5. It must take into account the economic aspect of life of the people to whom an educational institution belongs.

6. The curriculum should be realistic and rationalistic.

7. While forming the curriculum a balance be struck between the education of nature and education of man.

8. It should lay emphasis on learning to live rather than on living to learn.

9. In curriculum súch activities must be included, which help in preserving and transmitting the traditions knowledge and standards of conduct on which our civilisation depends.

10. It should be elastic and flexible.

11. It should be well integrated.

12. It should provide both for uniformity and variety.

13. It should be able to serve the needs of community.

As far as chemistry curriculum is concerned it should be elastic and variable, child-centred, community centred, activity centred. It should be such as to be use for adjustment in life and helps to integrate the activities of the child with his environment. It should be helpful to conserve and transmit the traditions, culture and civilisation. It must help in arousing the creative faculties of the children.

Planning Methods

There are a number of approaches to curriculum planning in chemistry. The extremes of such approaches are given in Table :

Table : The Extreme of Curriculum Formation

One extreme	*Other extreme*
Integrated	Disciplinary
Child-centred	Teacher-centred
Flexible	Structured
Process-based	Content-based
Conceptual	Factual

Actually no single way of curriculum planning exclusively based on one approach can fulfil the curricular needs of pupils. It is always better to combine different approaches to plan an effective curriculum in science.

Various Styles

Curriculum can be classified as:

(i) Instrumental curriculum.

(ii) Interactive curriculum.

(iii) Individualistic curriculum.

Instrumental Curriculum : In this type of curriculum more emphasis is placed on the utility value or vocational value of chemistry. It makes learning an intense competition among students.

The basic approach in such a curriculum is disciplinary and emphasises the acquisition of knowledge or information. The role of teacher is that of a dominant teacher in such a curriculum.

Interactive Curriculum : This type of curriculum is society oriented and lays more emphasis on the social development of child. In this type of curriculum class room instructions becomes an interactive or a cooperative process. The approach is interdisciplinary and the curriculum is 'loosely structured and consists of learning packages.

Individualistic Curriculum : In this type of curriculum more emphasis is placed on the personal development of the individual and it is based on interdisciplinary approach. It helps to develop creativity in the individual. This type of curriculum is based on self-calculation by the student.

Different Projects

In this section an attempt will be made to describe some of the chemistry curricula that have been developed over last thirty years or so. An attempt will also be made to give reasons for their introduction as also the way in which they were introduced.

The three early projects in chemistry were the following:

1. Chemical Bond Approach (C.B.A.) in United States.
2. Chemical Education Material Study (CHEM study) in United States.
3. Nuffield O-level chemistry in United Kingdom.

These projects influenced the mechanism for science curriculum reform in many countries through out 1960's and beyond.

Though there are a number of significant differences between the three projects cited above but they all arose at a time when a shortage of qualified scientific personnel was felt world-wide. Keeping in view the short comings of the existing curricula all these projects emphasised the following:

(i) Updating chemistry in the light of modern knowledge of the subject.

(ii) Giving the students a good understanding of the subject.

To achieve these ends the new curricula placed particular emphasis on such concepts as *periodicity* and *the mole.* They also incorporated some major chemical ideas underlying the *structure of materials, chemical bonding kinetics* and *energetics.* These are some tunes referred to as 'concept-based' which indicates the attention given to the principles of chemistry in their development. To make aware the students about the importance of chemistry topics like plastics, synthetic fibres, elastomers, detergents, drugs and insecticides were also included.

These curricula also emphasised the role of practicals (laboratory work) in chemistry which was seen as having a dual role. Firstly to illustrate and 'make real' the chemistry being taught and secondly to encourage scientific mode of thinking.

These projects were adopted by schools because of the participation of leading scientists like Glenn Seaborg (nobel prize winner) in united states and Sir Ronald Nyholm in United Kingdom.

Projects for Regions

In 1960's, in addition to national projects for curriculum development a number of projects were started to serve a large regional area consisting of several countries. One such project was the *UNESCO Pilot Project for Chemistry Teaching in Asia.* This project was aimed at bringing together chemical educators from various Asian countries in touch with one another and with their counterparts at other places in the world for the purpose of providing the necessary training in curriculum development. The well equipped laboratory at Bangkok in Thailand served as a regional meeting and working centre. The 'study groups' located in each Asian country provided information and consultancy services on innovations in chemistry teaching. The project lasted from 1964 to 1970.

Another regional project was the one which came to be known as the *school science project* in *East African Countries* of Kenya, Uganda and the United Republic of Tanzania. On the initiative of science teachers of these countries a British organisation then known as the Centre for Curriculum Renewal and Educational Development Overseas (CREDO) helped and G. Van Praagh ran courses for chemistry teachers. At a conference held in Nairobi in 1968, representatives from Uganda, Kenya and United Republic of Tanzania agreed to work together to produce new, 4- year courses in biology, chemistry and physics. These courses were intended to be up-to-date and relevant to the needs of the countries concerned. They were to be so designed as to stress understanding and for this purpose a substantial laboratory based component is to be incorporated in them. For curriculum preparation the ideas found in Nuffield Chemistry Project were extensively used. Drafts were prepared and tried in some schools and on the basis of feed-back they were revised. United Republic of Tanzania withdrew from the scheme in 1970. In Kenya and Uganda now a decision has been taken to fuse the traditional and newer courses into a single programme of study.

The project helped to raise the standard of awareness of and interest in, modern chemistry curricula in East Africa. It also helped to the publication of easily read background readers such as *salt in East Africa, Fermentation and Distillation*. CREDO played the role of coordinator.

The decision to choose between 'traditional' and 'new' curricula was left to schools. They may be considered as a good decision keeping in view the difficulties involved in preparing all teachers adequately and in a short-time, for large-scale science curriculum reform.

Projects for the Nation

Modern Chemistry Project in Malaysia is one such project. It not only concerns with development of modern curriculum but also concerns to help teachers to use it effectively, to improve the provision of laboratories and equipment and to produce a more appropriate form of examination for students who complete the course.

Another example is the nationally based chemistry project of Cuba. The new curriculum for schools in Cuba was developed with the assistance of specialists from USSR and the GDR. The new curricula is based upon two cycles, the first in grades 8 and 9 (two lessons per week) and the second in grades 10 to 12 (three lessons per week). In the first type, students study the principal types of inorganic compounds, their properties and general behaviour. They are also introduced to some fundamental chemical concepts and phenomenon. It includes teaching of the periodic law, electronic structure of atom and introduction to organic chemistry. The second cycle contains theory of electrolytic dissociation, energetics, chemical kinetics and chemical equilibria and organic compounds. The selection of content clearly illustrates the importance of Cuba's developing chemical industry.

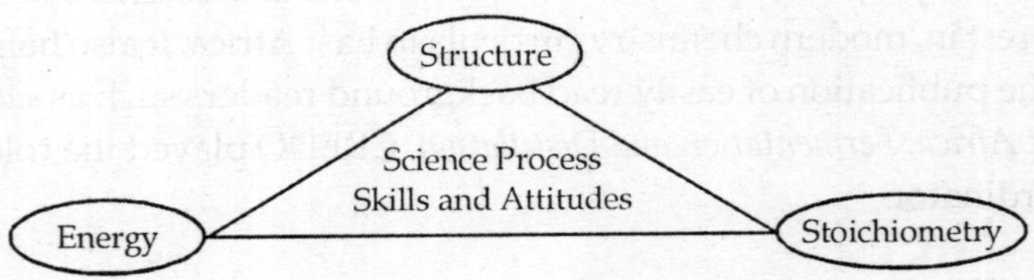

Fig. Theme for chemistry curricula, grades 11 and 12 (Thailand)

Projects of Advanced Status

Important advanced courses include those developed in Thailand and in India.

Thailand Project. New advanced chemistry course which is now in use in all secondary schools hi Thailand is built around the chemical themes illustrated in figure.

A frame work was built around these themes and then ideas and topics were shifted from one chapter to another during the process of development of curriculum.

There were many a difficulties in implementation of the new curriculum. However inspite various difficulties including those of finances the programme has been implemented across the whole of the Thailand. It has been well received and no need has been felt to make much revision in the course.

Indian Project. National Council of Educational Research and Training (NCERT) has developed a model senior-secondary level curriculum which has been adopted either as such or with slight modifications by a large number of states in India.

In a major change from traditional chemistry courses in India, in the new curriculum chemistry is presented as a unified subject. There is no traditional classification as physical, inorganic or organic chemistry. Some basic chemical concepts are developed in the beginning and these are later on applied during the study of elements and their compounds. An effort has also been made to relate macroscopic behaviour to microscopic behaviour. Two separate text books have been developed for classes XI and XII. Though some traditional open-ended and environmental investigations have been retained in the laboratory part of the course but the course is basically designed to develop skills, scientific attitudes and future training for research. Keeping in mind the interests of students who are likely to join vocational and professional courses an effort has been made to blend facts and theory by putting emphasis on the applications of the principles rather than their derivation. Equal emphasis has been given to chemical principles and descriptive chemistry.

Chemistry course at secondary level has also been revised. A review committee set up by the Ministry of Education to reduce work load, made drastic changes in the course recommending two alternative courses based on 'disciplinary approach' and a 'combined science approach'.

In some countries physics is more dominant in curriculum than chemistry.

In many countries, curriculum innovation in 1970's became increasingly involved with chemistry, not so much as a single subject, but as a part of a larger whole, chemistry now forms a part of courses in (i) physical science (ii) general science and (iii) integrated science.

Physical Science course adopted in some countries combines physics and chemistry courses together with omission of some subject matter.

General Science course combines the courses of physics, chemistry and biology. This type of course was based on the idea that general science should form an essential part of general education of all young people but unfortunately general science to meet the ideals.

Chemistry as a Component of Integrated Science

Some difficulty arises in the terminology in using 'General Science', 'science' and 'combined science' as all these have similar meaning. It may be clearly understood the 'integrated science' is in no way possesses a higher degree of integration than that possessed by science' or 'general science'.

Integrated science is generally more closely integrated usually with an element of social and environmental sciences. Most of the applications and social implications of science (*e.g.* Fertilisers; motor cars etc.) involve more than one science. Integrated science is widely associated with the movement to teach science for the majority. Making use of integrating theme 'Investigating the Earth' was an ambitions effort which exerted a considerable influence on the junior secondary curriculum in the United States. It includes structure, properties of matter and chemistry of the oceans, soils, minerals and atmosphere. This curriculum inspired many texts. Many countries are now adopting some form of integrated science, particularly in the early years of secondary school.

We can conclude the discussion by observing that despite disappointment with outcomes of some earlier projects, the 1980's brought a renewed commitment to science curriculum reform. In comparison to the earlier reforms the recent science curriculum initiatives identify the teacher or the teacher and learner as the focus of the reform so that curriculum improvement becomes essentially a matter of teacher development. Some curriculum initiatives have tried to accommodate 'metacognition' by fostering students' knowledge and awareness of, and thereby control over, their own learning.

The changed social and political context of 1980's has also had a marked effect upon school science education. Now we have courses which are concerned with science *e.g.* technology and society (STS) and the Chemical Education for the Public Understanding of Science Project (CEPUP).

However one trend has been remarkably constant, that towards laboratory work. Actually speaking, the school chemistry curriculum, in many countries, is now essentially laboratory based.

Conditions in other Countries

In China : In junior middle schools chemistry starts with a study of common things such as air and water. Then carbon, a very common element is introduced and also an important and common system of dispersal solution. The course concludes with the concepts of acids, bases and salts, the nature of an oxide and the rules of reactions.

In senior middle schools, the structure of matter and periodic table are introduced. These are based on the study of sulphur and alkali metals. From here they proceed to nitrogen and phosphorus. Other concepts included are rate of reaction, chemical equilibrium, electrolysis etc. The elements magnesium and silicon are also dealt with organic materials appear when oil industry and macromolecular compounds are simply explained.

In Germany: Students from class 7 to 10 are acquainted with fundamentals of structure of matter, the process of chemical change, important aspects of chemical production and applications of chemistry in other fields.

After their first few lessons on *structure of matter* students can differentiate between metal, molecular substances, ionic substances etc. from here they pass on to polymers.

At the end of the chemistry course the students are familiar with the fundamentals of chemical bonding, can differentiate between atomic bonds including polarised atomic bonds, ionic bonds and metallic bonds. They can relate the kind of bond to the property of the substance and are familiar with the structure and phenomenon of substances. They can also workout relationship between the structure and the reaction of organic substances. Chemistry imparts basic knowledge about major industries. Students also acquire knowledge about application of chemical science in other branches of production and spheres of life as well as about the inter-relationship between chemical industry and other industries. In the process they also realise the need of applying chemistry to well being of mankind.

Future Prospects

In future we are likely to move to more student-centred programme. For knowing the needs of various types of students (science and non-science) evaluation and research are essential. However the following general observations can be made for science and non-science students.

Science Students : Science students will be required to learn the qualitative and quantitative aspects of chemistry and so they should be given a rigorous course with mathematics through introductory calculus in secondary schools. This class of students is also expected to accept a more theoretical, more abstract approach in understanding chemistry. They require practical experiences in the laboratory and experiences in predicting, presenting data, designing experiments, measuring and evaluating results, redesigning etc. They should also be familiar with the method of using references, data books and library. They should be helped to develop the habit of scanning through text-books and journals and find ideas and answers, plan extension of experiments from the initial design provided to them.

The Non-Science Students : It has been estimated that only about 15-20 per cent of secondary school students are interested in science based carrier like engineer- ing, medicines, agriculture, nutrition etc. and the rest which forms the majority have other interests.

The knowledge of chemistry is highly needed even by these non-science students because if they understand advances in chemistry, the advantages and side effects of chemical industries in their nation their lives will be enriched.

There is an increase in literacy and specially in scientific literacy all across the world. In U.S.S.R., U.S.A., Japan and other developed countries about 80-90% of students get their science education which is compulsory upto Class X. The number of students studying chemistry is more than 70% of secondary school students and this number is more than 90% in case of U.S.S.R. The percentage of students undergoing science education is also on the increase in developing countries.

This increase in number of students undergoing chemistry courses is justified because a country cannot have a strong scientific and technological enterprise without a base in chemical education. Each nation needs creative scientists, engineers, science teachers and technologists. Each needs an informed citizen. Each nation needs individuals trained in the process of science and possessing an understanding of science to serve as school administrator, legislator and business and industrial leader.

At the international level, the degree of scientific literacy will help to determine the outcome of global questions related to food, population control, pollution, energy and peace.

Curriculum Contents

Secondary school chemistry in many countries is oriented to inorganic and physical chemistry. However students need an introduction to organic and biochemistry. Environmental chemistry must also be included in the curriculum of chemistry for secondary schools.

With introduction to organic chemistry we will be able to provide the students some understanding of petroleum chemistry, pharmaceutical chemistry and polymer chemistry which are essential areas of importance in economical growth.

Through biochemistry we will be able to introduce some important principles such as catalysis, rates of reaction, types of reactions, mechanism of reaction and chemistry related to human beings.

In environmental chemistry we can include such topics as chemical weathering, pollution, effects of fertilizers on the land and in the water etc. Some attention to colloidal chemistry and surface chemistry, nuclear chemistry will help to keep student interested in chemistry.

Curriculum Setup

We have already discussed various approaches such as chemistry as a part of general science, integrated science etc. for study of chemistry. Another view that has been discussed earlier is to have a *core syllabus* and an *optional syllabus*.

Another way of organising the curriculum is to design a two stream course for science students which should be mathematically rigorous, reasonably abstract and the second are a broad interdisciplinary chemistry based on integrating theme such as 'Investigating the chemistry of planet earth'. This may include section on cosmochemistry, geochemistry, biochemistry, environmental chemistry, nuclear chemistry etc.

The *set of modules* developed under the auspices of American Chemical Society (ACS) is the latest trend to organise chemistry curriculum. The module entitled *Combatting the Hydra* is designed for lower secondary schools. It consists of relevant units which acquaint students with the anticipated side effects for every new development. They use advances in fibres, post control, food additives and energy as examples. These small modules have been field tested and published for possible integration into general science curricula in United States.

Another project funded by National Science Foundation and directed by W.T. Lippincott of University of Arizona will develop a series of consumer-oriented chemistry modules to be class room tested and integrated into existing chemistry curricula.

The very fact that both these programmes are modular make it clear that it is almost impossible to replace existing curriculum and so an effort has to be made to develop material for insertion into existing curriculum.

Chemical Knowledge in Organised Form

The organisation of knowledge supports learning and retention of knowledge learned. Various theories have been propounded. Gagne's structuralist theory which is based on Blooms *Taxonomy of Educational* objectives has played a major role. *Science, A Process Approach* (SAPA) developed by American Association for the Advancement of Science used this theory to construct learning network of hierchies.

In recent publications written by educational psychologists we come across *cognitive maps* which facilitate learning by helping the learner to incorporate a large volume of information or concepts.

Reigeluth, Merrill and Bunderson recommend the structuring of subject-matter as the basis of deciding how to sequence and synthesise the modules of a subject-matter area.

The interest of chemists in structuring chemical knowledge has increased in recent years. Among the first who developed the structuring of chemical knowledge for chemical education is M. J. Frazer with his example for teaching of Faraday's laws. He presented a paper on this at the International symposium.

A number of other authors have tried to develop examples of structuring chemical knowledge. Basolo and Parry give examples of teaching systematic inorganic reaction without having students memorise specific individual reactions. Instead they should discuss these in terms of general reaction types *i.e.* of a system making extensive use of periodic table of elements. A reaction can be classified as a combination, decomposition, replacement, metathesis and neutralization.

Such structure reactions have existed in organic chemistry since long. Wilson gives an example of classification of electrophilic addition reactions of alkenes and alkynes. Patterns in organometallic chemistry with applications in organic synthesis have been discussed by Schwartz and Labinger.

Hall has proposed for an *organic chemists' periodic table*. Such a table be based on the movement of electrons and defining molecules as donors or acceptors. According to Hall, the periodic table of elements be considered as *Reactivity Map*, when plotting acceptor ability (abscissa) against donor ability (ordinate), the location of the resulting point gives an idea of the tendency for electron transfer to occur. Stronger donors be located down the table and stronger acceptors further across the table of *organic chemists' periodic table*. The more reactive they are, the more likely in the occurrence of transfer. However such a master table would become quite bulky and so construction of limited section of such a table in recommended.

Broad-based Status

Science Curriculum Programmes at national level began after the establishment of National Council of Educational Research and Training (NCERT) on 1st September 1961 with its headquarters at New Delhi. Besides many functions, NCERT develops curriculum, instructional materials, techniques of evaluation, teaching aids, kits and equipment etc.

In the last four decades, NCERT has developed several National Science Curriculum Programmes at all the school levels—primary, middle, secondary and senior secondary. In this chapter some of the several National Science Curriculum Programmes are discussed.

Programmes under UN

With a view to studying the existing science education programme in India a planning mission from UNESCO visited several States and Union Territories in 1964 and made some recommendations for improvement in teaching of science in Indian schools. In order to expedite the implementation of the scheme, it was decided by the Government of India, Ministry of Education and Social Welfare to launch a pilot project from the beginning of the next academic year. The pilot project covered Primary classes (I-V) and Middle classes (VI-VIII), and was funded by UNICEF. This project was called UNICEF-Assisted Science Education Programme (SEP).

Discipline in Practice

Under this project NCERT developed a national primary science programme, "SCIENCE IS DOING" in 1970. Its adapted, adopted and translated (in regional languages) versions were used in most of the primary schools in the country. This programme was a package of:

(a) Class III-V Text Books;

(b) Class III-V Teacher's Guides;

(c) Primary Science Kit and Kit Guide;

(d) Class I-II Syllabus;

(e) Two 16 mm films for Teacher Training;

(i) Science is Doing; and (ii) Primary Science Kit.

In spite of such a good primary science programme when we went into the classrooms, we found that science was not doing; science was either reading, telling or in very few cases 'science was demonstrating.

Research in Effect

In Delhi Directorate of Education, Siddiqis conducted research studies on cognitive development of primary school children, during 1975-77 on 1206 Delhi Primary school children. It was found that majority of primary school children (95.6 per cent) is either pre-operational or concrete operational, and a very small percentage (4.4 per cent) is at formal operational stage. Similar results could also be obtained in other parts of the country. This shows that working with concrete objects or doing experiments is a very important part of primary science education.

The following recommendations are based on these studies for the purpose of providing guidelines to teachers who teach science to primary school children and to science educators who develop instructional material—textbooks, teachers' guides, audio-visual aids etc., and design teaching techniques:

1. Discourage traditional teaching techniques like book reading and teacher telling.

2. Involve children in activities with concrete objects and doing experiments with their own hands.

3. Encourage children to find out facts of science by doing experiments and not to memories.

4. Delete the science concepts and skills from the existing science texts which are not compatible with the cognitive level of children.

5. Delete the activities from the existing science texts which are not compatible with the cognitive level of children.

6. Introduce only those concepts which are compatible with the cognitive level of children of a particular class and in which use of concrete objects in their immediate environment may be possible.

7. Use local resources and environment which is full of real and concrete objects when teaching science to primary school children.

Similar studies could also be conducted on children at other school levels and the results could possibly have a positive effect on science teaching at all levels. These studies reveal that if our primary school teachers used environment and local resources and teach science through environmental approach, the learning will be more meaningful.

Science of Environment

The National Policy on Education (Kothari Education Commission Report 1964-66) adopted by the Government of India recommends that science should form an integral part of general education for the first ten years of schooling. Therefore, NCERT examined the question of what type of science courses should be introduced at various levels of school system of 10+2 pattern. One of the major recommendations of the NCERT is that the child should learn the method of inquiry in science and should begin to appreciate science and technology in the life and the world around him. The NCERT recommends further that in classes I and II science should be taught as Environmental Studies (EVS), which includes both the natural and the social environment. Later on in classes III, IV and V, two subjects, *viz.*, EVS (social science) and EVS (general science) should be taught. There are certain areas where there will be variation depending upon the environment of the child. Hence, it is proposed that the equipment and materials locally available should be used for the teaching of science. Cycle, bullock cart, motor pump etc., available in the village set-up can be used in explaining many important concepts in science. Nature Should be used as a laboratory for science teaching.

On the basis of this (the Kothari Commission Report) NCERT had laid down certain objectives for teaching science through environment. They are stated as follows:

1. The main objective is to enable children to observe their environment and to enrich their experience, thereby developing skills in the processes of science, such as observing, communicating, measuring, hypothesising, and experimenting to test the hypotheses.

2. Besides developing skills in some processes of science, the children, through EVS (General Science) if given knowledge of scientific facts and principles, have a better understanding of the phenomena taking place in the environment around them.

3. The understanding of the environment through application of scientific method (the processes of science) will help children to develop a scientific attitude in life, which may comprise such components as rational outlook, open-mindedness, a positive inclination for democratic, secular, and socialistic outlook to situation in life, opposition to the prejudices based on sex, caste, religion, language or region.

4. Children will be helped to develop their creative faculties—their imagination and independent thinking for locating problems, suggesting solutions and trying out their ideas.

Curriculum for Initial Stage

'Science is Doing' syllabus for classes I-V was revised. Environment which is full of concrete objects was taken into consideration when the revised science syllabus was framed, and the instructional materials were developed based on the revised Environment Studies (EVS) syllabus. The new curriculum, 'EVS Programme' developed by NCERT has the following materials:

(i) EVS class I-II (science and social studies) Teacher's Guide;

(ii) EVS (science) textbooks for class III, IV and V.

This programme is to be adopted or adapted by the States and Union Territories depending upon their own environment and local resources.

UNICEF also funded the interested States and Union Territories for developing 'Handbook of Activities Using Environment and Local Resources' for the use of primary school teachers teaching science, so that they could use NCERT developed EVS textbooks more effectively when teaching science with environmental approach.

Science Branch, Directorate of Education, Delhi also took this project. According to the findings of the Siddiqis' Research studies on the "Cognitive Development of Primary School Children (1975-77)" majority of primary school children (95.6 per cent) are either pre-operational or concrete operational and they need concrete objects to learn science concepts and skills. Our environment is full of real and concrete objects, which the children can use to learn science. Based on the findings of these research studies and their implications a "Handbook of Activities Using Environment and Local Resources" was developed by Science Branch, Directorate of Education, Delhi for the use of primary school teachers teaching science and social studies to classes I and II and science to classes III, IV and V. This UNICEF assisted programme consists of an Environmental kit and a package of seveit booklets:

(i) Class I Activities;

(ii) Class II Activities;

(iii) Class III Activities;

(iv) Class IV Activities;

(v) Class V Activities;

(vi) Environmental Kit Guide;

(vii) Class I-II (science and social studies) and classes III-V (science) syllabus.

This is the bank of activities for use of teachers. For one minor idea several activities have been developed to be fit in different environments. This programme was tried out and revised according to the feedback received from experimental schools. Its English version is also available which might be useful for those who are interested or engaged in developing such materials in other States and Union Territories of the country as well as in other countries.

Upto early sixties science at middle level was a general science course. It had a little content and in it no importance was given to science processes. Learning science was a book reading process with no emphasis on experimentation and demonstration. Only once a while some very innovative teachers used to do some demonstrations when teaching science. This process continued till 1964 when UNESCO came into picture, and a new disciplined science course came into existence for middle classes.

This was a package of physics, chemistry and biology courses at the middle stage. These courses were introductory in nature and helped the students to familiarise with the basic concepts and processes compatible with the cognitive levels of students at this stage.

The main objectives of the science course were:

(i) to develop scientific knowledge;

(ii) to develop processes of science such as observation, experimentation, problem-solving and investigatory approach;

(iii) to develop scientific attitude;

(iv) to develop mechanical, experimental and mental skills;

(v) to develop the interest of science in children; and

(vi) to appreciate the role of science in everyday life.

Based on these objectives NCERT developed disciplined science courses in physics, chemistry and biology in mid-sixties. These courses were oriented in such a way that teaching of science was based on first hand experiences, practical experiments, which might be in the form of demonstrations by the teacher and individual laboratory exercises and up-to-date facts of science with suitable examples and illustrations from everyday life.

Physics Course. The physics course was a 3-year programme (VI, VII and VIII). The main objectives of physics teaching were:

To enable the child to:

(i) obtain the knowledge and understanding of the basic concepts and laws of physics;

(ii) acquaint themselves with the application of laws in everyday life;

(iii) obtain the understanding of the physical phenomena and basic laws governing them;

(iv) develop skills in solving problems, handling equipment and performing experiments;

(v) develop scientific attitudes and the spirit of scientific enquiry;

(vi) appreciate the laws of physical science in everyday life.

Based on these objectives, the physics programmes developed by NCERT was a package of:

(a) Class VI-VIII Physics textbooks;

(b) Class VI-VIII Physics teachers' guides;

(c) Class VI-VIII Physics test items;

(d) Class VI-VIII Physics kits (3—one for each class) and kit guides;

(e) teacher training films (Physics kit—Parts I, II and III).

Chemistry Course. The Chemistry course was a 2-year programme (VII and VIII). The main objectives of Chemistry Teaching were:

To provide the children with:

(i) knowledge of the nature of substances, their properties and their atom-molecular composition;

(ii) knowledge and understanding of the nature of chemical changes, their types and basic laws governing them;

(iii) development of power of observation and ability to explain the chemical phenomena involved in the processes studied;

(iv) acquaintance with language of chemistry;

(v) development of skills in solving problems, handling substances and equipments to do experiments;

(vi) development of computational skills;

(vii) development of scientific attitude and spirit of scientific enquiry;

(viii) application of the role of chemistry in everyday life.

Based on the objectives, the chemistry programme developed by NCERT was a package of:

(a) Class VII-VIII Chemistry textbooks;

(b) Class VII-VIII Chemistry teachers' guides;

(c) Class VII-VIII Chemistry test items;

(d) Class VII-VIII Chemistry kit and kit guide;

(e) Teacher training films (Chemistry Demonstration Kit).

Biology Course. The Biology course was a 3-year programme (VI, VII and VIII). The main objectives of Biology Teaching were:

(i) To acquaint the students with the world of living things which surround them;

(ii) To discover the main laws and regularities governing the life processes;

(iii) To give an understanding of the nature of biographical science;

(iv) To create interest in pupils in life sciences enabling them in the solution of problems of everyday life;

(v) To give knowledge of:

(a) plants and their functions (Botany);

(b) animals and their functions (Zoology);

(c) the structure and functions of the human body (Human Physiology);

(d) man and his environment.

Based on these objectives, the biology programme developed by NCERT was a package of:

(a) Class VI-VIII Biology textbooks;

(b) Class VI-VIII Biology teachers' guides;

(c) Class VI-VIII Biology test items;

(d) Class VI-VIII Biology kit and kit guide;

(e) Teacher training film (Biology kit—Parts I, II and III). The Disciplined Science textbooks were written in such a way that science teachers in schools would have physics, chemistry and biology kits, but they could not be supplied to all the schools. Thus again, science teaching became a book reading or teacher telling process. Disciplined science curricula developed during sixties were assessed as inappropriate for the eighties. Isolated disciplines served to cut science off from our common humanity. For this it was decided to seek a curriculum with more attention to integrative system and operations that can interweave diverse disciplines.

Projects of Different Status

During seventies a number of integrated science teaching projects and programmes were developed in various countries. These programmes embody a wide range of different approaches to integrate science teaching, including processes, concept, units, environment, applied science, thematic projects and patterns. As a part of 10+2 curriculum the NCERT also developed an integrated science programme for middle classes (VI-VIII).

If we review the overall aims and objectives of integrated science programmes, we find these three aspects are reflected, *i.e.,*

(a) the nature of science; (b) the nature of learners, and (c) the nature of society.

The Nature of Science

(i) The simple cognitive skills, such as the ability to recall knowledge, comprehend the basic concepts and themes of science such as the nature of matter, energy and its transformations, the nature and properties of living things, etc., and their application in novel situations.

(ii) The process skills such as the ability to observe, measure, classify and predict. It is these skills that will largely influence the way in which the subject matter is taught.

(iii) The development of attitudes such as honesty and open-mindedness, and the realisation of the tentative nature of science theories. These are unlikely to arise from the courses unless the approach to practical investigation rejects true query.

(iv) The skills appropriate to science: these would include not only the ability to manipulate apparatus, but also to construct and interpret tables, charts and graphs, as well as to be able to find relevant information from sources of reference.

The Nature of Learners. It seems that many integrated science teaching programmes, most of which cover a period of several years of schooling, attempt to reflect the findings of research in the development of science and mathematics concepts in children. It looks, there appears to be considerable awareness in the designers of integrated science schemes that curriculum development should be geared to the thought of the child and not just to be logical structure of the subject.

The Nature of Society

(i) In the selection of aims for the course as a whole greater attention is paid to the overall needs of society. One reason for the widespread popularity of integrated science courses is an awareness that they can better reflect the aspirations of society than courses in single science discipline.

(ii) In the selection of subject matter, the social significance of science has been given prominence and technology has also crept into integrated science courses. Thus, the Integrated Science Project places heavy emphasis on integration of science, technology and society.

Based on these objectives a full curriculum package containing syllabus, textbooks, teachers' guides, kit and kit guide was developed during 1975-80. The process of development includes participation and interaction of more than a hundred people belonging to the category of scientists, subject teachers, method masters, science educators and classroom teachers. All worked to fulfil certain predetermined objectives drawn on the basis of the curriculum frame work prepared by NCERT in 1975.

This integrated science curriculum was finally implemented. Some states have already introduced it and many are contemplating its introduction. In the States and organisations where this curriculum has been introduced, it has been observed that the textbooks are being used as if these are mere amalgamation of physics, chemistry and biology. The curriculum is not being implemented in the intended spirit.

Integrated science teaching covers all those approaches to teaching science:

(a) in which concepts and principles of science (physics, chemistry, biology, etc.) are presented in such a way as to express to fundamental unity of scientific thought,

(b) which emphasis the processes and methodology of the scientific outlook, and

(c) which embody a scientific study of the environment and technological recruitments for everyday life. The NCERT integrated science curriculum can be analysed keeping an eye over the above given definition of integrated science.

Programmes for Schools

The Education Commission headed by Professor D.S. Kothari, was appointed by the Government of India in July 1964. It submitted its report in July 1966, recommending guiding principles and working policies for the development of Indian Education at all stages and in all respects. Based on its recommendations 10+2+3 Education Scheme started in Delhi Schools, Central Schools and other schools under Central Board of Secondary Education (CBSE) throughout the country in July 1975, in which science was compulsory (as a part of general education) like other subjects upto class X. The first Secondary School Examination was conducted by CBSE in 1977. The same year the new Science Course was introduced at +2 stage in class XI in Senior Secondary Schools, and in 1979 the first batch of class XII students appeared in Senior Secondary Examination conducted by CBSE. NCERT developed secondary and senior secondary science curricula were used in the schools under CBSE.

This programme developed by NCERT in 1975 was a package of three textbooks—physics, chemistry and life science. These textbooks after being written were placed before a group of selected science teachers for review. Only after minor changes these textbooks were in the market for the use of students.

This secondary science course lived for four years and in 1979 it split up into two courses—Science A-Course and Science B-Course, according to the recommendations of Ishwarbhai Patel Review Committee.

A-Course. The syllabus of the programme was framed by CBSE, and the books were written by private publishers. This was a package of the following nine books:

(i) Physics Theory Part I for Class IX

(ii) Physics Theory Part II for Class X

(iii) Physics Practical for Classes IX and X

(iv) Chemistry Theory Part I for Class IX

(v) Chemistry Theory Part II for Class X

(vi) Chemistry Practical for Classes IX and X

(vii) Life Science Theory Part I for Class IX

(viii) Life Science Theory Part II for Class X

(ix) Life Science Practical for Classes IX and X.

B-Course. This course was developed by NCERT in 1979. This was a package of two textbooks:

(i) Science Part I for Class IX

(ii) Science Part II for Class X

For science practical, private publishers published science practical books (Physics, Chemistry and Life Science all in one) based on CBSE prescribed syllabus.

These NCERT developed science textbooks were written on integrated approach, and not as separate textbook of physics, chemistry and life science, as it is clear from the sequence of chapters in the two textbooks.

Science Part I (for Class IX)

1. Our Universe
2. Living World: An Introduction
3. Motion
4. Moments and Couples
5. Work and Energy
6. Atomic and Molecular Masses, Mole Concept and Chemical Equation
7. Behaviour of Gases
8. Flotation
9. Elasticity of Solids
10. The Structure of the Atom
11. Chemical Bonding
12. Oxidation and Reduction
13. Periodic Classification of Elements
14. The Halogens
15. Oxygen and Sulphur
16. Nitrogen and Phosphorus
17. Organisation of Life
18. Man and his Environment

Science Part II (for Class X)

19. Wave Motion
20. Reflection and Refraction of Light
21. Carbon
22. Compounds of Carbon—Organic Chemistry
23. Life Processes
24. Genetics and Evolution
25. Electricity
26. Magnetism
27. Solutions and Electrolytic Dissociation
28. The Rates of Reactions and Chemical Equilibrium
29. Combustion and Fuels
30. Human Biology, Health and Nutrition
31. Metals and Metallurgical Processes
32. Some Applications of Science.

If we look at these two science courses, we find that Science A-Course consists of a vast syllabus as compared to Science B-Course. Science A-Course includes nine books (three in each Physics, Chemistry and Life Science), while in Science B-Course all the three subjects (Physics, Chemistry and Life Science) are amalgamated in one book (in two parts), plus book for science practical.

According to the Ishwarbhai Patel Review Committee on the curriculum for the ten year school, published by the Ministry of Education and Social Welfare (1978) the aim of framing Science A-Course was "to develop special interest in students for the science subjects," while Science B-Course aimed at imparting in students a broad based knowledge of science along with other subjects.

This means that the study of Science A-Course demanded an intensive teaching and teachers are expected to put in some more work, and show extra interest in the teaching of this course. But the number of periods allotted per week (usually 9, each of duration 30-40 minutes) to the teaching of Science A-Course remained the same as are allotted to the teaching of Science B-Course. There were only 3 periods per week allotted to each subject physics, chemistry and life science (theory and practical).

Under these conditions it was proposed that in order to make Science A-Course more useful and popular among the students the following recommendations (the first two of the Review Committee, 1978) may be considered by the curriculum developers and educational administrators:

1. The main concepts should be studied and unnecessary details which overload the syllabus should be avoided.

2. The content of the course should be capable of being taught or studied within the time allotted.

3. The Science A-Course should cover the pre-requisites of the senior secondary (XI-XII) science.

Science for All

One of the recommendations of the Education Commission under the chairmanship of Dr. D.S. Kothari was that the science should be made a compulsory subject in school education.

The recommendation was accepted and science was made compulsory up to class X in several States and Union Territories, creating a lot of problems. In the old 3 years Higher Secondary Scheme (IX-XI), a selected number of students (about 33 per cent) who were really interested in science used to take science in class IX. But now like other subjects science is also a required course upto class X. Therefore all students including a big majority of those who perhaps do not have any interest and aptitude for science have to take science. Now the question arises, must every student be forced to learn the same science for which he has no aptitude and interest or may a special curriculum in science be developed for such students so that science may be interesting even for them.

Introduction of Courses

The Science A and B Course started in 1979 finished in 1984 after living for five years. The Central Board of Secondary Education, New Delhi started one science course from 1984. It is 3 in 1—Physics, Chemistry and Biology.

This programme was developed by NCERT in *1977-78*, starting with textbooks in Physics, Chemistry and Biology for Senior Secondary (XI-XII) students. Afterwards teachers' guides and test-items were also written on these texts. This programme was introduced at +2 stage in class XI in Senior Secondary Schools in July *1977*, and in 1979 the first batch of class XII students appeared in Senior Secondary Examination conducted by CBSE.

After visiting hundreds of schools, observing science classes, interviewing science teachers, students, principals and parents some key problems regarding science education at +2 stage have been identified. They are as follows:

1. There seems to be a big gap between senior secondary science (Physics, Chemistry and Biology) and secondary science courses.

2. It looks that there is no co-ordination between theory and practicals in science Specially in physics at +2 stage.

3. The science laboratories at +2 stage are not equipped enough with all the needed facilities for all the experiments, investigatory projects as well as demonstration experiments.

4. Students, teachers, administrators and parents seem to be not satisfied with the science courses at +2 stage.

5. Science courses at +2 stage do not fulfil the basic admission test requirements of some professional courses like engineering and medicine, as students seeking admission to these courses need extra content coaching.

6. At +2 stage in class XII besides experiments in Physics, Chemistry and Biology students are also to work on one investigatory science project (in each Physics, Chemistry and Biology). It looks that many science teachers are not clear as to what the difference between a science project and investigatory science project is, and they need orientation in guiding their students on some investigatory science projects.

Senior Secondary Science courses were revised afterwards but solutions to such problems (mentioned above) were not taken into consideration satisfactorily when revising these courses.

The responsibility of drawing up science curricula, writing of textbooks and designing of teaching aids for the introduction of science in schools has been given to NCERT. The curricula has been drawn up and textbooks written by those who perhaps have never taught in schools (Indian Express, May 5, 1981). They have tried to compress in the textbooks, a variety of topics some of which have little utility or relevance. What students upto the secondary stage (in general education) should learn, is the application of science to everyday life and not theoretical science. Better science perhaps was taught before as Raman, Saha, and Bhabha were products of pre-independence science. In spite of the Ishwarbhai Patel Committee Report on Curricula Reform for the ten years school submitted in 1978, the curricula have not been updated. Even the Minister of Education presiding over the annual general meeting of the NCERT on March 11, 1981 regretted that attempts at improving the quality of education in terms of upgrading curricula were not heartening.

Morarji Desai, the then Prime Minister, while addressing the members of the Ishwarbhai Patel Review Committee on the curriculum for the ten year school at New Delhi stated, "the books that I did carry in college are being carried by school students today. The students are burdened by how many books, I do not know."

Knowledge of science doubles every decade. We are to keep pace with the new development. The problem is how much knowledge in science should be given to a child at a particular level, so that he is not burdened.

Is the capacity of a child to absorb knowledge limited? Professor V.N. Wanchoo, President, All India Science Teachers Association, in his Presidential Address in the 22nd All India Science Teachers' Conference at Trivandrum in December 1978, has tried to answer this question:

"Research on theory of learning and the development of brain during the last two decades indicate that a child is capable of learning any amount of knowledge at his level. What is really lacking in this country, are the management and transmission technique of information and the process and the channels through which such information should be imparted at different school levels. The exploration of knowledge is affecting all the countries uniformly. Others deviate their attention to the problems of management of information, and we appoint committees which end up by making some marginal changes in the syllabus. Deleting an information in one class and adding an item of information in another class, will not serve any purpose; and this is what exactly Ishwarbhai Patel Committee has done in recommending changes in the Ten-Year School Curriculum proposed by NCERT".

QUESTIONS

1. (a) Describe 'Science is Doing' programme?

   (b) How is EVS programme different from 'Science is Doing'?

2. 'Handbook of Activities Using Environment and Local Resources' is based on the findings of the Research Studies on the Cognitive Development of Primary School Children. Discuss.

3. What is the difference between:

   (a) Disciplined Science Programme, and

   (b) Integrated Science Programme? Discuss.

4. What is the difference between:

   (a) Science A-Course and

   (b) Science B-Course. Which course do you think will be a better course as a prerequisite of senior secondary science? Discuss.

5. What are some key problems regarding science education at +2 stage? What do you think will be the possible solutions? Discuss.

15

Lesson Planning

Though a syllabus is prescribed for each class yet the teacher is at liberty to draw up his own teaching syllabus. It is best to organise the teaching syllabus around a few broad areas of experience of pupils. For this purpose the syllabus is divided into a number of units.

Planning of Unit

A *unit* is a related learning segment made up of a few lessons along with an outline of its actual execution in the class room. Thus a unit will consist of both the subject matter and methodology of its delivery to students.

Hoover defines unit as, "The teaching unit is a group of related concepts from which a given set of instructional and educational experiences is desired. Unit normally range for three to six weeks long".

In view of Preston, a unit is a large chunk or a block of related subject matter as can be over viewed by the learner.

After having divided the prescribed syllabus into a number of teaching units the teachers will decide the time that could be allotted to each unit. After that he can break up each unit in a number of lessons and each lesson should be complete in itself. After this the teacher will enter in his diary the scheme of work under various headings.

Some of the advantages of unit planning are as under:

(i) It provides a basic course structure around which specific class activities can be organised.

(ii) It enables the teacher to integrate the basic course concepts and those of related areas into various teaching experiences.

(iii) It provides an opportunity to the teacher to keep a balance between various dimensions of the prescribed course.

(iv) It enables the teacher to break away from traditional textbook teaching.

Unit No.................................

Date	*Course content*	*Demonstration*	*Equipment material*	*Student's activities*	*Remarks/ References*

If the prescribed course has to be course has to be covered in a number of years then it is unwise to distribute the course in units spread over a number of years.

Unit Planning Proforma for Chemistry

Grade Level...

Unit Level...

Behavioural Objection......................................

(i) ....

(ii) ....

(iii) ....

Sr. No.	*Major concepts from the content*	*Number of periods and lessons*	*Teaching Method to be used*	*Teaching aids to be used*
1.				
2.				
3.				
4.				
5.				

Planning of Lessons

A proper planning of the lessons is key to effective teaching. The teacher must know in advance the subject matter and mode of its delivery in the class room. This gives the teacher an idea of how to develop the key concepts and how to correlate them to real life situations and how to conclude the lesson. Lesson planning is also essential because effective learning takes place only if the subject matter is presented in an integrated and correlated manner and is related to the pupil's environment. Though lesson planning requires a hard work but it is rewarding too. L.B. Stands conceives a lesson as 'plan of action' implemented by the teacher in the class room. According to G.H. Green, "The teacher who has planned his lesson wisely related to his topic and to his class room without any anxiety, ready to embark with confidence upon a job he understands and prepared to carry it to a workmanable conclusion. He has foreseen the difficulties that are likely to arise, and prepared himself to deal with them. He knows the aims that his lesson is intended to fulfill, and he has marshalled his own resources for the purpose. And because he is free of anxiety, he will be able coolly to estimate the value of his work as the lesson proceeds, equally aware of failure and success and prepared to learn from both".

The Advantages : Some of the advantages of planning a lesson are as under:

(i) Lesson planning makes the work regular, organised and more systematic.

(ii) It induces confidence in the teacher.

(iii) It makes teacher quite conscious of the aim which makes him conscious of attitudes he wants to develop in his students.

(iv) It saves a lot of time.

(v) It helps in making correlation between the concepts with the pupils environment.

(vi) It stimulates the teacher to ask striking questions.

(vii) It provides more freedom in teaching.

Main Features : Some Important features of a good lesson plan are as under:

Objectives: All the congnitive objectives that are intended to be fulfilled should be listed in the lesson plan.

Content: The subject matter that is intended to be covered should be limited to prescribed time. The matter must be interesting and it should be related to pupil's previous knowledge. It should also be related to daily life situations.

Method(s): The most appropriate method be chosen by the teacher. The method chosen should be suitable to the subject matter to be taught. Suitable teaching aids must also be identified by the teacher. Teacher may also use supplementary aids to make his lesson more effective.

Evaluation: Teacher must evaluate his lesson to find the extent to which he has achieved the aim of his lesson. Evaluation can be done even by recapitulation of subject matter through suitable questions.

Formal steps in lesson planning are:

(i) Introduction (or preparation)

(ii) Presentation

(iii) Association (or Comparison)

(iv) Generalisation

(v) Application

(vi) Recapitulation

Introduction : It pertains to preparing and motivating children to the lesson content by linking it to the previous knowledge of the student, by arousing curiosity of the children and by making an appeal to their senses. This prepares the child's mind to receive new knowledge. This step though so important must be brief. It may involve testing of previous knowledge of the child. Some times the curiosity of pupil can be aroused by some experiment, chart, model, story or even by some useful discussion.

Presentation : It involves the stating of the object of lesson and exposure of students to new information. The actual lesson begins and both teacher and students participate. Teacher should make use of different teaching aids to make his lesson effective. Teacher should draw as much as is possible from the students making use of judicious questions. In chemistry lesson it is desirable that a heuristic atmosphere prevails in the class.

Association : It is always desirable that new ideas or knowledge be associated to the daily situations by citing suitable examples and by drawing comparisons with the related concepts. This step is all the more important when we are establishing principles or generalising definitions.

Generalisation : In chemistry lessons generally the learning material leads to certain generalisation leading to establishment of certain formulaes, principles or laws. An effort be made that the students draw the conclusions themselves. Teacher should guide the students only if their generalization is either incomplete or irrelevant.

Application : In this step of lesson plan the knowledge gained is applied to certain situations. This step is in confirmity with the general desire of the students to make use of generalisation in order to see for themselves if the generalisations are valid in certain situations or not? No lesson of chemistry may be considered complete if such rules, principles, formula etc. are not applied to life situations.

Recapitulation : In this last step of his lesson plan the teacher tries to ascertain whether his students have understood and grasped the subject matter or not. This is used for assessing the effectiveness of the lesson by asking students questions on the contents of the lesson. Recaptulation can also be done by giving a short objective type test to the class or even by asking the students to label some unlabelled sketch.

One most important point to remember is that the six steps given above for lesson planning should not try to follow these very rigidly. These are only guide lines and in many a lessons it is not possible to follow all these steps.

There is another way of lesson planning which is gaining currency these days. It is known as *Glover Plan*. This plan has four steps as follows:

Questioning. Teacher must introduce and develop his lesson through related and sequential questions. Start the lesson by asking questions about previous knowledge of the students. The questions should then lead to new knowledge under consideration.

Lesson can also be introduced with the help of some teaching aid like a picture, chart or model etc. the introduction can also to made by describing a situation or by telling a short story.

However teacher should hear in mind that the introduction is brief and interesting.

Discussion. For discussion the class be divided into smaller groups and in such groups students be encouraged to express their ideas and opinions freely. This helps the students in removal of their difficulties.

Investigation. The students are encouraged to do a project or investigation on the lesson topic either individually or in small groups by processing information or by laboratory work.

Expression. It concerns the strategy in which the student's and teacher's communication of ideas through observation and listening (passive expression) or through doing (active expression) or through fine and performing arts (artistic expression) or by arranging learning situations (organisational expression).

In developing a lesson a teachers must keep in mind the following psychological principles.

Principle of Selection and Division. The teacher should wisely select and divide the learning material into smaller segments. It is also for the teacher to decide about the quantum of subject matter to be covered by him and that which has to be illicited from the students.

Principle of Successive Clarity. It is for the teacher to see that the different learning segments of lesson are well structured, sequenced and connected. Teacher must ensure, at each segment, that students have grasped the subject matter given to them.

Principle of Integration. Teacher should conclude his lesson only after combining various learning segments to produce some generalisation.

Lesson Plan Designing

Lecture-cum-Demonstration Method : The style given below is generally followed for writing a lesson plan.

Class: Date:

Subject: Duration:

Topic: of period:

Instructional Matrial ....

....

General Objectives ....

Specific (Objectives ....

....

Previous Knowledge...

....

Questions

1. ....?
2. ....?
3. ....?

Introduction

Questions

1. ....?

2. ....?

Announcement of Aim ....

....

Presentation

Matter	*Method*	*B.B.Summary*

Generalisations ....

....

Applications ....

....

Recapitulation ....

....

Questions

1. ....?

2. ....?

3. ....?

House Task ....

....

Model Lesson Plan

Class-X Date:

Subject—Chemistry Duration: 40 minutes

Topic—Composition of Air

Instructional Materials

1. Chalk board, duster, coloured chalks.
2. Trough, jar, match box, phosphorus etc.
3. Candle, glass tumbler, house hold plate, baby feeder etc.

General Objectives

1. To develop scientific attitude amongst the pupil.
2. To develop lower of observation and sense of enquiry amongst the pupil,
3. To develop reflective thinking in the pupils.

Presentation

	Matter	*Method*	*Black-board Summary*
1.	Yellow phosphorus in air at 307K.	Showing phosphorus, teacher asks what is this? Does phosphorus burn in air?	Yellow Phosphorus It is kept under water. It burns in air.
2.	Yellow phosphorus is kept in water.	What is yellow phosphorus stored water?	
3.	Yellow phosphorus burns in air.	What happens if a piece of yellow phosphorus is kept in air?	
4.	Fitting up the apparatus for the experiment as shown in fig.	What is this? What is this? What is this?	
5.	Phosphorus piece is allowed to burn by touching it with hot iron rod.	What happens? Why does phosphorus burn? What is this cloud like substance?	
6.	Water rises upto mark No. 1.	Why has water risen up in the bell jar?	
7.	Phosphorus pentoxide is soluble in water.	What does it signify? Where has phosphorus pentoxide gone?	
8.	No more oxygen is present in the bell-jar now.	Absence of oxygen in the bell-jar can be tested by taking in burning match stick inside the bell-jar.	Air contains one part of oxygen and four parts of nitrogen. It is soluble in water.

Specific Objective

To tell the students that air contains one part of oxygen and four parts of nitrogen by volume.

Previous Knowledge

It is presumed that students know that air contains oxygen and nitrogen. They also know that oxygen is a supporter of combustion and that a burning candle goes out in nitrogen.

Introduction

To introduce the lesson teacher will pick up a coin in his fist and will ask the following questions while taking away the coin,

1. What is in my fist? (A coin)
2. When the coin has been taken away? What is now in my hand? (Air)
3. It is possible for us to live without air? (No)
4. Name the gases present in Air? (Oxygen, Nitrogen and some CO_2, inert gases etc.).
5. What is the proportion of oxygen and nitrogen in the air?

Announcement of Aim

On out receiving a proper reply to question 5 teacher will announce the aim "Today we will try to know about the proportion of oxygen and nitrogen in air".

Generalisation

From the above experiments we conclude that oxygen and nitrogen are present, in the ratio of 1 : 4 by volume, in air.

Recapitulation

Teacher will ask the following questions for recapitulation. Does phosphorus burn if exposed to air?

Home Task

Students will be asked to perform a similar experiment using a candle instead of phosphorus.

A Few Examples

Sodium and its Chief Compounds

Aim: To teach the physical and chemical properties of the metal sodium, and the names and common uses of some of its important salts.

Previous Knowledge: Students know the distinctive features of metals. They are also familiar with the names and know the common uses of washing soda, caustic soda and common salt.

1st Stage: Introduction. Following questions will be asked to test previous knowledge:

(i) Name the chief characteristics of metals.

(ii) Give some important properties that distinguish metals from non-metals?

(iii) Name an element which though lighter than water is yet a metal. Why do you suppose it to be a metal?

(iv) To what use do we put caustic soda, washing soda and common salt?

Teacher will then declare the aim: 'We shall learn more about the metal sodium and substances like caustic soda, washing soda, etc. today'.

2nd Stage: Some Properties of Sodium. Teacher will put a freshly cut piece of sodium on a filter paper and pass it round the class to show its metallic lustre; a student will be asked to cut another piece and its soft waxy nature will be brought home. Similarly its lightness and the effect of exposure to air will be shown and reasons for storing it under kerosene oil will be explained.

Blackboard. Sodium is a light, soft metal. When freshly cut it shows a metallic lustre. When exposed to air it soon gets tarnished. It floats on water and soon disappears, so it is kept under kerosene oil.

3rd Stage: Action of Sodium on Water. Experiment, to show that hydrogen is given out when sodium react with water, and that an alkali is also formed which turns red litmus solution blue, will be shown. A glass tubing of slightly wide bore will be supported in a beaker containing red litmus solution. One or two small pieces of sodium will be dropped inside the tube. The gas coming out of the tube will be ignited with a match. The litmus solution inside the tube will be seen to have turned blue.

Blackboard. When a piece of sodium is thrown into water it swims around with a hissing sound, reacts with water, giving hydrogen and forming an alkali which turns red litmus blue.

4th Stage: Some Common Compounds of Sodium and their Uses. Samples of common salt, caustic soda, washing soda, and sodium bicarbonate will be shown, and the class will be told that all of them are compounds of metal sodium. Students will be asked some of the uses of these salts and other uses will be told to them. Their chemical names will also be given to the students.

Blackboard

(i) Common salt—sodium chloride is used for (a) eating, (b) curing hides and fish, (c) preservative in achars and other things (d) in the making of washing soda and caustic soda, and preparation of hydrochloric acid.

(ii) Washing soda—sodium carbonate is used for (a) washing, (b) softening hard water and (c) making caustic soda.

(iii) Sodium bicarbonate is used in (a) medicine and (b) baking powders.

(iv) Caustic soda—sodium hydroxide is used in making soap and paper.

5th Stage: Recapitulation.

(i) Why is sodium not stored under water or in an empty bottle?

(ii) Give some physical and chemical properties of sodium.

(iii) List the uses of common salt, soda and caustic soda.

Blackboard Summary. Main properties, uses etc. as above will form B.B. Summary.

Preparation and Study of the Chief Properties of Carbon Dioxide Gas in the Laboratory

Aim: To get pupils to fit up the apparatus for the preparation of carbon dioxide in the laboratory, prepare the gas and study its chief physical and chemical properties.

Previous Knowledge: The preparation and properties of the gas has been demonstrated in demonstration period.

Procedure: The students will be asked the following questions:

(i) How was carbon dioxide prepared in the laboratory?

(ii) Was it heavier or lighter than air?

(iii) Was it soluble in water?

(iv) How can you collect the gas?

A sketch of the apparatus will be drawn on the blackboard and teacher will ask them to fit the apparatus accordingly.

The following precautions will also be emphasized:

(i) The apparatus should be tested to be air-tight.

(ii) The thistle funnel should dip in the liquid.

(iii) Water should be just enough to cover the pieces of marble.

(These will be written on blackboard)

The boys will then be asked to fit up the apparatus. The teacher will go round giving individual help. While the boys are busy collecting the gas, he will put down on the blackboard a list of experiments to be performed and properties to be tested by the boys as given below:

(i) Colour and smell.

(ii) Action on litmus solution.

(iii) Heaviness.

(iv) Action on a burning taper.

(v) Action on lime water for a short-time and for a long-time.

He will ask them to record their work in the following tabular form:

Experiment	***Observation***	***Inference***

When the boys are busy performing the experiments to study the properties of gas, the teacher will go round, give individual help and tick off portions of written work examined.

At the end of the period he will sign the notebooks and supervise the return of clean apparatus to the cupboards.

Note: In the specimens of lesson notes given above, the apparatus required has not been shown. Pupil teachers should always give in their notes the list of apparatus required. This list may be put below the aim, under the heading 'apparatus'.

16

Judging the Capability

It has long been recognised that for curriculum development to be successful, the assessment of students must be sensitive to the aims and objectives of curriculum. It judiciously employed assessment results can be used to evaluate curricula, particularly to determine difficulties. *Evaluation* is a new term in the field of education that has been introduced to replace the terms like testing or examination etc.

Evaluation has a wider meaning as compared to testing or examination. Concept of testing is very much limited in terms of objectives, scope, methodology etc. where as evaluation has a very wide meaning as it includes to assess all educational outcomes and outputs which have been brought about by teaching-learning process. Recent trends in learning and evaluation link them to behavioural objectives specified for a course of study in chemistry. Actually a total change in behaviour of the learner related with all the three domains (conative, cognitive and affective)is expected by learning experiences provided to him.

In this chapter an attempt will be made to study the specific procedures for evaluating the effectiveness of chemistry teaching-learning.

Efforts for Betterment

The sense of discontentment with the prevailing system of examination, in India, can be easily traced back to British days. A report submitted by Zakir Hussain Committee in 1938 recommended for longer duration test so as to cover the whole of the curriculum. The examination be given in such a form that would make marking objective and independent of individual judgement.

The examination committee of the Central Advisory Board of Education gave its report on, "Post War Examination Developments in India" in 1944 and recommended as under, ".....every attempt should be made to devise and standardise objective-type tests for use in this country so that they may supplement and ultimately replace the old type of examinations".

These recommendations were never implemented and they remained on paper only.

After attaining independence in 1947, proper attention was given to examination reforms. Radhakrishanan Commission (The first Education Commission) on university education (1949) reported as under to bring to the fore the weaknesses essay-type examination prevailing in our universities.

"An unsound examination system Continuous to dominate instructions to the detriment of a quickly expandinag system of education. In our visits to universities we heard from teachers and students alike, the tale of how examinations have become the aim and end of education, how all instructions is subordinated to them, how they kill initiative in the teacher and the student, how capricious, invalid, unreliable and inadequate they are and how they tend to corrupt the moral standards of university life".

".....we are convinced that if we are to suggest one single reform in university education, it should be that of examination".

The Secondary Education Commission (1953) also recommended a reform in system of examinations. In this report we find, "In order to reduce the element of subjectivity of essay-type tests, objective tests of attainment should be widely introduced side by side. Moreover the nature of the tests and type of questions should be thoroughly changed. They should be such as to discourage cramming and encourage intelligent understanding".

Another commission commonly known as Kothari Commission (1966) made the following remarks in its report, about reforms in examination system.

".....But the task is a stupendous one, and it will take considerable time for new measures to make their impact on objectives, learning experiences and evaluation procedures in schools education".

The commission made many recommendations for lower primary, middle and other examinations.

Examination reforms have also been advocated with National Policy on Education (1968). It states, "A major goal of examination reform should be to improve the reliability and validity of examinations and to make evaluation a continuous process aimed at helping the student to improve his level of achievement rather than at 'certifying' the quality of his performance at a given moment of time".

The Measurement

In words of Kothari Commission (1966) "Evaluation is a continuous process, it forms an integral part of the total system of education, and is intimately related to educational objectives. It exercises a great influence on the pupil's study habits and teachers' methods of instruction and thus helps not only to measure educational achievement but also to improve it the techniques of evaluation are means of collecting evidences about the students development in desirable directions".

Evaluation, thus may work as a connecting bridge between the objectives of teaching science and the ways and means of attaining these objectives in the form of learning experiences, learning methods and learning environment.

A students' learning is evaluated in terms of the extent of achievement and then behavioural objectives specified for a course of study in chemistry. Behavioural objectives are specific, observable and measurable aim and serve as a guide for learning and are desired for the eventual achievement of a general objective.

Relationship among Objectives, Learning Experiences and Evaluation : The close relationship that exists between objectives, learning experiences and evaluation is depicted in Fig.

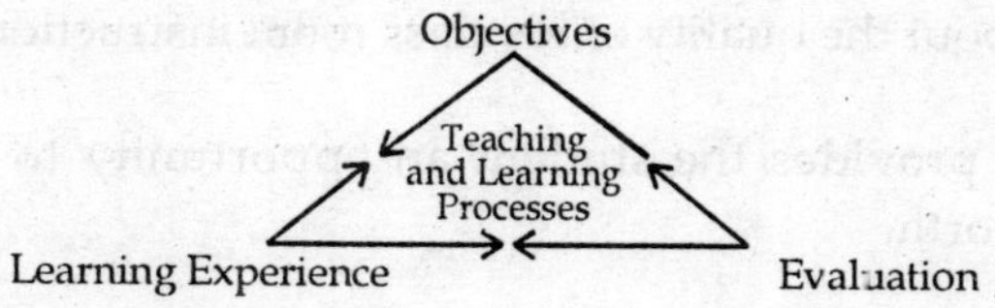

Fig.

The learning experiences for any topic in a subject are designed keeping in view the study of that topic. These learning experiences are likely to bring about behavioural changes in the learner as specified through different objectives. Evaluation of students' performance is generally done in terms of marks or grades competitively. Sometimes students may be compared with some absolute performance standard instead of making comparison with other students of a given group. Thus there are two kinds of evaluation (i) criterion refernced evaluation and (ii) Nonrefernced evaluation.

Criterion-Referenced Evaluation : It assesses the students performance in term of a specified performance standard or criterion without any mention of the performance levels of the other students of the group. This evaluation method is related to mastery and developmental tests.

Norm-Referenced Evaluation : It assessed the students performance relative to other students of the group. Students are awarded marks and relative ranks in this method of evaluation.

Objective of Tests

Evaluation fulfills the following purposes:

(i) It assesses the extent of learning by students and gives them the feed-back about their performance.

(ii) It gives feed-back to the teacher about the learning gaps of the students. It also provides the teacher a feed-back about the quality of his class room instructions.

(iii) It provides the student an opportunity to show his worth.

(iv) It serves as a screening tool for selecting students for special purposes.

Our evaluation has another goal besides assisting the teacher in assessing and modifying her teaching procedures. This goal *self-evaluation* is not solely for the students. As teacher and students actively engage in all levels of a study such as initial planning, organizing and carrying out activities they can be guided in developing ability to evaluate themselves. Knowing the general and specific goals can aid the pupil in checking himself all along the way. This makes the learner an active participant in class room activities. It also places same of the responsibility on him for learning and assessing what and how much he has learned. Self guided evaluation stimulates healthy and realistic achievement goals. A logical first step self evaluations is setting up of realistic goals. These goals for chemistry teaching in elementary schools are:

(i) *Functional understandings* such as concepts, principles, generalizations, and the facts needed.

(ii) *Problem Solving Skills* such as defining problems, proposing hypothesis and techniques necessary for the solution of the problems, observational techniques, discussion and interpretations skills.

(iii) *Scientific attitudes, interests and appreciation* such as open mindedness and humanity.

The easiest area to evaluate is functional understanding because a rich variety of tests are well known and are widely used in elementary schools. Before we proceed to actual discussion of these tests let us consider the criterion of a good examination and pre-requisites of a physical test.

Examination Standards

Though a variety of tests are available to test the functional understanding of the child but for true assessment of such aspects of growth as the elements of reflective thinking, scientific attitudes, resourcefulness, creativeness or such other objectives or interests we require more precise and accurate instruments of evaluation. According to most of the psychologists and educationalists the following are essential criteria of satisfactory evaluation.

Validity. Any good test should measure what it claims to measure.

Reliability. A good test is one that is reliable *i.e.* it gives same rating to a candidate even if he is examined by different examines and even at different times.

Pre-requisites of a Good Test

There are certain pre-requisites for preparing a good test. These are as under:

Aspects	*Description*
Aims	— Acquisition of knowledge of various concepts and skills.
	— Development of scientific attitude and interest.
	— Development of laboratory skills.
	— Highlighting the application of chemistry in every day life and technology.
	— Development of skills of information processing, observation, enquiry and design.
	— Acquisition of problem-solving abilities.
Knowledge :	Recall and recognition of factual information such as:
	(i) definitions of various terms.
	(ii) statement of laws, principles, rules, conventions etc.
	(iii) Description of construction and working of devices and instruments.
	(iv) Description of events, processes and phenomenon.
	(v) Recognising the parts of devices, instruments, appliances and apparatus.
	(vi) Identifying known physical phenomenon, events and occurrences.
Comprehension :	Understanding facts, laws etc.
	(i) Comparing and contrasting various phenomenon.
	(ii) Locating errors, limitations and defects.
	(iii) Illustrating scientific phenomenon.
	(iv) Reasoning events on the basis of scientific principles and laws.

Contd.

Aspects		***Description***
Applications :		Using knowledge in various situations.
	(i)	Solving numerical problems.
	(ii)	Making use of various scientific laws in various situations and events.
	(iii)	Relating various scientific variables.
Skills :		Using psycho-motor skills.
	(i)	Laying out an experimental set-up.
	(ii)	Drawing diagrams, graphs, histograms, flow charts etc.
	(iii)	Reading various measuring instruments.
Analysis :		Breaking up information into parts to reach conclusions:
	(i)	Interpretation of observations.
	(ii)	Drawing inferences from observations.
	(iii)	Generalising conclusions.
Synthesis :		Combining parts of information to grasp a concept.
	(i)	Designing an experiment.
	(ii)	Improvising and experiment, apparatus or device.
	(iii)	Improving the accuracy of an instrument.

Objectivity. A test can be considered objective if the scoring of the test is not affected in any way by the examiner's personal judgment. Thus the opinion, bias or judgment of the examiner can have no influence on the results of an objective test.

Comprehensiveness. By comprehensiveness of a test we mean that it covers the whole or nearly the whole course content and the questions are uniformly distributed to cover the course content.

Practicability. A test is called practicable if it can be easily administered and is acceptable to average examines. While preparing such a test, the time and cost of administration must be taken into consideration. The test should be usable and should serve a definite need in the situation in which it is used.

Interpretibility. A lest can be considered as interpretable if its scores can be used and interpreted in terms of a common base having natural or accepted meaning.

Easy to Administer. A good test should be easy to administer so definite provision be made for collection and preparation of test material. It should give simple, clear and precise instructions.

Table : Allocation of Marks for Abilities to be Tested

Ability	***Symbol***	***Marks***
Knowledge	K	45
Comprehension	C	26
Application	A	17
Skills	S	6
Analysis and Synthesis	An/Sn	6
Total		**100**

Test Designing

A good test should be constructed in accordance with a definite design or plan. The steps in designing a test are as under:

(i) Allocation of marks for the different cognitive levels to be tested.

(ii) Allocation of marks for different chapters or units.

(iii) Blue print for the question paper.

(iv) Allocation of marks to various types of questions.

After the blue-print is ready the actual question paper is set. Some of the commonly used tests in chemistry are fill-ins, true-false, multiple-choice, short-answer or essay-type etc.

Now we shall take up the discussion of some of these tests.

Functional Skills under Observation

Concepts, Generalization and Principles : The need of written tests becomes increasingly important as children progresses through the elementary school grades. This is so because of the following reasons:

(i) In upper grades pressures for more "objective evaluation" in chemistry are greater as children are exposed to greater emphasis upon "subject matter grades".

(ii) As children's use of language increases, there can reasonably be greater emphasis upon meaningful written and verbal concept development.

(iii) As the child builds a background of chemistry concepts, facts, understandings and inter-relationships, a greater need is presented for accurately assessing the child's knowledge.

(iv) With larger classes, as is generally the rule for the intermediate and upper grades, teachers require evaluation techniques that are fast, accurate, and easy to apply, score and interpret.

One of the types of written testing devices is the *short-answer tests*. One major disadvantage is the superficiality and isolation of factual materials asked for rather than a breath and depth of understanding. They do however offer the teacher:

(i) opportunities for including wide ranges of items to be tested.

(ii) an ease of writing questions because of the shortness of each.

(iii) a minimum of time and effort is needed for scoring because of the shortness of answers expected, and

(iv) opportunities for involvement of pupils in the self-evaluation because of the ease of scoring and following up incorporate responses.

Basically there are two types of short answer testing devices recall and recognition examinations.

Recall Tests

As the term implies, recall questions ask the student to bring back to mind information that the student was exposed to in the past. Psychologists have indicated that the people usually associate items to be recalled with other items and information and rarely, if ever, completely isolate them. The way in which individuals associate isolated items is still much of a mystery. Even tests of isolation such as the inkblot design used in *Rorschach test*, evoke widely divergent responses because of unique back grounds and associations of individuals. Recall with children thus becomes a problem of framing questions in such a way as to stimulate the remembrance of the situation in which the intended information occurred. One of the ways in which this can be accomplished on recall tests is formulation of a question so that only one word or a few words is needed to answer the query. This simple question and answer procedure might look like this.

What is the approximate %age of oxygen in air at sea level?

Another way of accomplishing recall of information in a chemistry content study is by supplying statements with blanks to be filled in.

For Example

Two by-products of the process of photosynthesis are .... and ....

Recognition Tests

True and False tests are probably the most commonly used recognition tests in use today. The basic idea involved is illustrated as follows.

True	False

Such tests encourage guessing and it greatly reduces the validity and reliability of the Tests. Because it is very difficult to frame questions that are neither too obvious nor too ambiguous, this type of examination should be used very sparingly. Whenever possible other types of recognition tests such as multiple choice test should be given.

Multiple Choice Test

In this type of test, several alternatives are presented to the pupil from which he must select the one that makes the statement most correct. Such test items can reduce the subjectivity in marking and inter-examiner variability in marking. These tests are the most popular these days and are most useful because in this way guessing is minimised and intelligent thinking is encouraged. Some examples of this type of tests are:

A scientist who studies rocks is called (a) a chemist *(b)* a geologist (c) a geometer *(d)* a geopolitist.

Reasoning power can play a big part in answering this type of questions and so called educated guesses should be encouraged. Actually these educated guesses usually are formulated from vague relationships that are seen or sensed. Very often the person cannot explain his reason for selection of correct choices in this type of questions, he just knows. Because there are so many aspects of learning and teaching that are still mysteries to us, teacher should not stand in the way of children learning. Intuition plays an important part in learning as well as in the scientific way of working.

Guidelines for Constructing Multiple-Choice Test Items : While constructing multiple choice test items following guidelines be followed:

1. A test-item should have a single concept to be tested.

2. A test-item should be such that it can be used to discriminate a group of students as low, medium and high achievers.

3. The statement of test-item should be very clear and unambiguous.

4. Be sure that of the plausible answers only one answer is correct.

Parts of Multiple Choice Item : There are generally two parts of a multiple choice test-item, *viz.*, stem and plausible answers. The stem of the test-item contains the statement of the question or problem. There are some important styles of writing the stem of a multiple choice questions. These are:

1. Stating the stem in the form of a question.
2. Writing the stem as an incomplete statement.
3. Writing the stem as a problem to be solved.

The plausible answers are the options available to the student from which he has to choose the correct answer. These are generally written according to following guide lines:

1. Write the answers in such a way that to a student who has not read the topic thoroughly each answer seems to be plausible.
2. Include common misconceptions which an average student holds about a particular learning segment.
3. Options which are true on their own but defy the statement of the problem given in the stem of the test item.
4. Do not provide clues for the right answers.

Cognitive Levels and Multiple Choice Items : Generally at the school level, the multiple choice questions in chemistry are related to three cognitive levels, *viz.*, knowledge, comprehension and application.

The Limitation : Some of the limitations of objective type tests are:

(i) They fail to test the ability to organise material.

(ii) They cannot test how well a thought is expressed.

(iii) They encourage guess work.

(iv) They are difficult to design.

Test by Matching

Besides the true and false and the multiple choice tests, there is a third type of recognition test, *the matching lest*.

In this type of test items two mismatched columns are given, one working as problem statement and the other working as options. The questions and answers given in two columns are required to be matched or compared by the students. By giving the pupil two columns of items and asking him to match the related items, the teacher can quickly and easily see if his student recognises the relationships that exist between the items. There is less of a stress upon sheer memory or recall of fragmentary information because the materials are presented to the student for his correlation.

Because matching tests are focused mainly to measuring subject matter, it is not always indicative of the pupils ability to perceive the deeper meaning or real understanding of the relationship between the items used on the tests. Stress upon mere verbalization and memory of isolated bits of information should be avoided. Teachers will find it necessary to use all types of testing instruments so as to get a broad picture of the formulation of his children's chemistry concepts.

Short Answer System

With all the drawbacks of the short answer tests, there is a wide use for these tests in chemical education in schools. They are becoming quite popular these days. As the name suggests such questions expect brief, to the point, limited short answers. Generally the length of answers is specified. They offer the teacher an ease of construction and scoring not possible with other types of tests. The tests offer a greater degree of objectivity than other evaluating techniques and the results of tests can be helpful to the teacher for evaluating and reporting children's progress in chemistry education to their parents. With the teacher's guidance, the simplicity of tests can be useful for self-evaluative examinations for the children. Children can also be involved in writing examinations of this type as well as in scoring them. Teachers can be assured that the objective tests being discussed warrant the expenditure of time and effort required to construct them in correct way. Correctly made, administered and interpreted, the short answer test offers many advantages to the teacher; however, they should never be used as sole testing device. They should only be used in conjunction with other types of oral and written tests as well as teacher observation.

Advantages of Short Answer Questions : Some of the important advantages of this type of questions are:

(i) They are easy to design.

(ii) Scoring is less subjective and easy.

(iii) The question paper becomes comprehensive i.e. it covers the entire syllabus. The students lose the chance of spotting questions or topics.

Short Answer and Structured Questions : Perhaps the most interesting development over the past twenty years or so is that of the structured questions of the following type:

5.0 cm of 2.0 m aqueous solution of Y chloride (where Y is a metal) was placed in each of eight similar test tubes. Different volumes of 2.0 m aqueous solution of silver nitrate were then added to the solution in each of the test tubes. The resulting mixtures were shaken and allowed to settle. The heights of the precipitates obtained in each test-tube were plotted against volumes of the silver nitrate solution added. The graph obtained is shown in Fig.

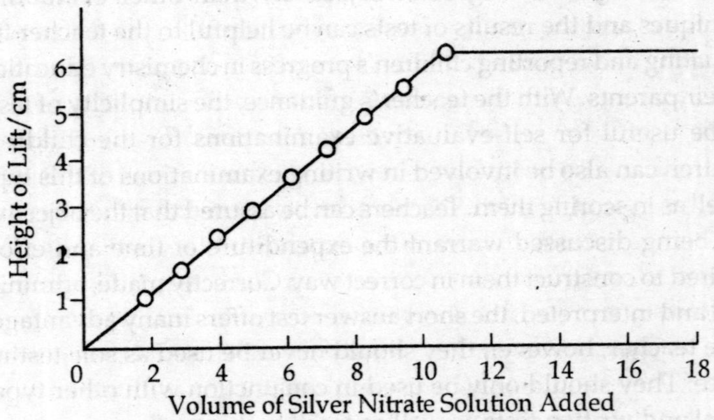

Fig.

(a) (i) Name the precipitate formed.

(ii) What is the initial colour of the precipitate?

(b) How many moles of Y chloride is present in 5 cm^3 of 2.0 m solution of Y chloride?

(c) What is the volume of silver nitrate solution that will be just sufficient to react completely with 5.0 cm3 of Y chloride solution?

(d) Calculate the number of moles of silver nitrate that will react with one mole of Y chloride.

(e) If the volume of aqueous solution of Y chloride used in slightly more than 5.0 cm^3, the maximum height of the precipitate obtained will be different. Sketch the graph you would expect to obtain on Fig.

(f) 14.0 cm^3 of silver nitrate solution is mixed with 5.0 cm^3 of Y chloride solution. The precipitate formed in filtered.

(i) What will be observed if the precipitate obtained is exposed to sun light for a few hours?

(ii) Metal Y is above copper in the electro-chemical series, describe what will be observed when a piece of copper foil is placed in the filtrate.

This involves the art of *questioning*, questioning so structured and phrased as to stimulate a response from most of the students and thus lead them to the understanding of the whole.

A *structured question* is essentially one in which the student is asked to study information given in the stem, usually complex and generally unfamiliar, and is then asked to respond to it through a series of questions, each requiring a short answer.

Though the use of such structured questions is on increase but training in their construction is not so readily available. Teacher can use the following guidelines for constructing such questions guidelines given below also summarise the form of this type of assessment.

Guidelines for Framing Structured Questions

1. The *stem* should provide information and act as a focus for the set of questions which follow it.

2. The questions following the stem should relate to the stem. These questions should be in a sequence according to one or more of the following principles:

   (a) a teaching sequence through which the information normally would be studied.

(b) a logical sequence of operations such as steps of a calculation.

(c) increasing difficulty a hierarchy of skills.

3. In formulation of *questions* due thought be given to expected *responses*. For a precise answer the question too should be *precise*.

4. Marks allotted to each question be indicated against each.

5. Enough space be provided, between questions, for expected answers.

6. Normally five or six question are framed from a stem.

7. Discretion be allowed to examiners for marking answers to such questions.

8. Generally choice is not allowed.

9. A set of question may have either a linear structure or a branched structure. The branched type is more common in chemistry. In this type the questions do not depend on one another although they all relate to common stem.

10. Like objective tests, structured questions can be used for formative evaluation and diagnosis.

Tests by Essay

The essay-type examinations are in use in India since long and these have been greatly appreciated due to the freedom of response allowed. Essay tests aid in evaluating chemistry understanding in the intermediate and upper grade of elementary school. Like all testing devices, essays present many serious disadvantages.

At the same time they present many possibilities for gathering informations. This type of test directs attentions to and places emphasis on a larger segment of the subject or on an integrated total unit. It provides the student a chance to create a new approach to a problem as it requires the student to express his views in writing. He is required to produce some thing and not merely to guess or recognise the answer. These tests can measure verbal fluency, skill of expression, organisation of thoughts and the attitude of examinees towards problems and subjects considered in the class, however it lacks most of the qualities of a good measuring instrument.

Obvious advantages and disadvantages of essay test.

1. Shows how well the student is able to organise and present ideas

   but

   scoring is very subjective due to a lack of set answers.

2. Varying degrees of correctness since there is not just a right or wrong answer,

   but

   scoring requires excessive time.

3. Tests ability to analyze problems using pertinent information and to arrive at generalization or conclusions,

   but

   Scoring in influenced by spelling, handwriting, sentence structure and other extraneous items.

4. Gets to deeper meanings and inter-relationships rather than isolated bits of factual materials,

   but

   Questions usually are ambiguous or too obvious.

The Limitation : These tests have low validity, low reliability and are less comprehensive. Discussing about the weaknesses of this type of tests Ross remarks, "The essay overrates the importance of knowing how to say a thing and under-rates the importance of having some thing to say".

This type of tests are not reliable because there is no agreement between teachers about the marks to be assigned and studies have shown that even the same teachers do not agree with themselves. Sandifoard, in his book on educational psychology refers to a study, "In one department of the University of Toronto, the same subject was set for an essay in different years. The essay which had secured 80 marks in one year, was exactly copied by the students in another year and scored 39 marks".

Ashburn who carried out a study at University of West Virginia concluded that, "the passing or failing of about 40% depends not on what they know or do not know, but on who reads the papers and that the passing or failing of about 10% depends on when the papers are read".

Another general complaint of students about essay type tests is that the questions 'did not suit them'. Certainly nine or ten questions generally set in this type of question paper cannot cover the whole syllabus. Hence this types of test is less comprehensive.

If an effort to offset the disadvantages the teacher must carefully consider the construction of each essay question. The teacher should word the question in such a fashion that the pupil will be limited to a certain degree to the concepts being tested.

To minimise the shortcomings of excessive subjectivity teacher should prepare a scoring guide before hand. Each question be scored separately and a list of important ideas that are expected should be made.

Practical Work Assessed

Assessment of practical work is the most difficult operational problem in assessment. The reason for it may that curriculum designers and teachers are not clear in their mind about the objectives to be achieved. In part the major aim of practicals was the mastery of manipulative skill, presently there are many other aims. In one study twenty one aims have been given out.

For assessment of practical work there are three alternatives before us:

(i) The assessment be done by external examiners.

(ii) Internal assessment system may be followed.

(iii) Practical work may not be assessed at all.

Presently most of the practical assessment work is done by external examiners. However this form of assessment of practical work has the following disadvantages:

(i) A large number of students have to be examined simultaneously.

(ii) A large number of similar sets of apparatus, equipment etc. are required for this type of assessment.

(iii) Reliability of single practical examination is suspect.

However inspite of its various shortcomings this system of practical examination is in use because it is thought that any practical examination is better than none.

On a limited scale internal assessment of practicals has been undertaken. This type of assessment is based on the belief that assessment of practical work of students by their own teacher on several occassions during the course of study shall be more reliable than one single examination by external examiners. Source of the advantages of internal assessment of practical work are:

(i) The reliability increases because of increase in frequency of examinations.

(ii) In this system the range of attributes of students is extended and it includes those which are displayed during work as well as at the end of it.

(iii) In this type of assessment range of experiments and types of work can be extended.

However the dual role of teacher in such type of assessment may sometimes result in adversely affecting the relationship between the teacher and the taught.

As a safeguard to such a system of assessment are may take recourse to moderating the scores by source external moderator. But this is a lengthy and cumbersome process. Some other statistical methods such as moderation on the basis of some written examination can also be undertaken.

However it can be easily seen that internal assessment of practical work involves both teachers and administrators in a good deal of work and it demands a high level of competence and professional integrity.

Project Work Assessed

"In project work, pupils are expected to assume some level of personal responsibility for their work and to organize their time for constructive study".

Project Categories

There important types of project are:

Report Type. In this type of project students collect information from books, journals and other sources and then prepare a report in the form of a project report.

Discovery Type. In this type of project the students use the results of their own experimentation, observation etc. to answer a specific question of a specific hypothesis. These findings are then summarised as a project report.

Combination Type. In this type the theoretical and experimental aspects of a topic are combined and thus is actually a combination of report type and discovery type of project.

The objective of project work is to develop the skills of planning design, investigation and interpretation.

The assessment of project work becomes difficult because of such tall claims for projects as an educational activity, more over most of the chemistry projects are of co-operative nature and so in such cases it would be difficult to differentiate between the performances and attributes of members of a group is self evident.

This clearly brings about the problems in assessment of projects. However to have same type of project work and same type of its assessment is considered better to the rejection of all project work in chemical education just because it can not be properly assessed.

Summary : If we compare the present with the past (20-30 years ago) we find that these days much more thought and resources go into assessment of attainment of chemistry. The use of microprocessors is on the increase which is likely to lighten the burden of a cumbersome test processing and administration system. It is also likely to facilitate international test banking and application.

An inexpensive electronic information source of large capacity is now available which can be readily used by students during a chemistry examination. It opens up the possibility of a revolutionary open book type of assessment very shortly.